Building MATH SKILLS

STEP BY STEP

Intermediate Book B

Dr. Annette L. Rich

Illustrated by Debora Weber

ISBN 0-8454-2438-6

Copyright © 1990 The Continental Press, Inc.

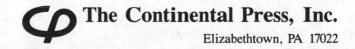 **The Continental Press, Inc.**

Elizabethtown, PA 17022

Contents

Name _____

2 + 2 + 2 = _____
3 twos = _____

3 + 3 + 3 + 3 = _____
4 threes = _____

4 + 4 = _____
2 fours = _____

5 + 5 + 5 = _____
3 fives = _____

You can add to find how many in all. An easier way to find how many in all is to multiply.
The × sign means multiply.

2 + 2 + 2 + 2 = _____
4 twos = _____
4 × 2 = _____

5 + 5 = _____
2 fives = _____
2 × 5 = _____

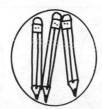 

3 + 3 = _____
2 threes = _____
2 × 3 = _____

4 + 4 + 4 + 4 = _____
4 fours = _____
4 × 4 = _____

Multiplication as Related to Addition
© The Continental Press, Inc.

3

Name _____

5 fours = _20_
5 times 4 = _20_
5 × 4 = _20_

3 threes = _____
3 times 3 = _____
3 × 3 = _____

2 twos = _____
2 times 2 = _____
2 × 2 = _____

4 fives = _____
4 times 5 = _____
4 × 5 = _____

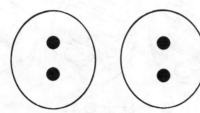

5 threes = _____
5 times 3 = _____
5 × 3 = _____

4 fours = _____
4 times 4 = _____
4 × 4 = _____

2 threes = _____
2 times 3 = _____
2 × 3 = _____

5 twos = _____
5 times 2 = _____
5 × 2 = _____

4

Name _____

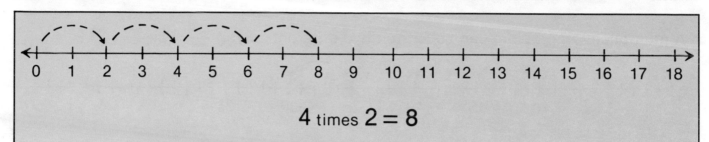

4 times 2 = 8

4 × 2 = 8

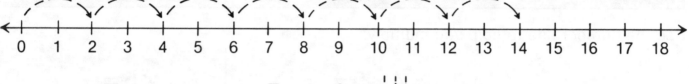

7 times 2 = _14_

7 × 2 = _14_

Use the number line to help you multiply.

1 × 2 = ____	4 × 2 = ____	7 × 2 = ____
2 × 2 = ____	5 × 2 = ____	8 × 2 = ____
3 × 2 = ____	6 × 2 = ____	9 × 2 = ____

Multiply.

4 × 2 = ____	7 × 2 = ____	5 × 2 = ____
6 × 2 = ____	3 × 2 = ____	1 × 2 = ____
9 × 2 = ____	8 × 2 = ____	2 × 2 = ____
2 × 2 = ____	4 × 2 = ____	7 × 2 = ____
5 × 2 = ____	6 × 2 = ____	9 × 2 = ____
8 × 2 = ____	1 × 2 = ____	3 × 2 = ____

Multiplying 2
© The Continental Press, Inc.

5

A multiplication problem can be written two different ways.

0   5   10   15   20   25   30   35   40   45   50

$$6 \times 5 = 30 \qquad \begin{array}{r} 5 \\ \times 6 \\ \hline 30 \end{array}$$

Use the number line to help you multiply.

$1 \times 5 = \underline{5}$ $\qquad$ $4 \times 5 = \underline{}$ $\qquad$ $7 \times 5 = \underline{}$

$2 \times 5 = \underline{}$ $\qquad$ $5 \times 5 = \underline{}$ $\qquad$ $8 \times 5 = \underline{}$

$3 \times 5 = \underline{}$ $\qquad$ $6 \times 5 = \underline{}$ $\qquad$ $9 \times 5 = \underline{}$

Multiply.

$$\begin{array}{r} 5 \\ \times 4 \\ \hline 20 \end{array} \quad \begin{array}{r} 5 \\ \times 7 \\ \hline \end{array} \quad \begin{array}{r} 5 \\ \times 9 \\ \hline \end{array} \quad \begin{array}{r} 5 \\ \times 5 \\ \hline \end{array} \quad \begin{array}{r} 5 \\ \times 2 \\ \hline \end{array} \quad \begin{array}{r} 5 \\ \times 1 \\ \hline \end{array} \quad \begin{array}{r} 5 \\ \times 3 \\ \hline \end{array} \quad \begin{array}{r} 5 \\ \times 6 \\ \hline \end{array}$$

$$\begin{array}{r} 5 \\ \times 2 \\ \hline \end{array} \quad \begin{array}{r} 5 \\ \times 8 \\ \hline \end{array} \quad \begin{array}{r} 5 \\ \times 3 \\ \hline \end{array} \quad \begin{array}{r} 5 \\ \times 9 \\ \hline \end{array} \quad \begin{array}{r} 5 \\ \times 6 \\ \hline \end{array} \quad \begin{array}{r} 5 \\ \times 5 \\ \hline \end{array} \quad \begin{array}{r} 5 \\ \times 7 \\ \hline \end{array} \quad \begin{array}{r} 5 \\ \times 8 \\ \hline \end{array}$$

Multiplying 5

Name _____

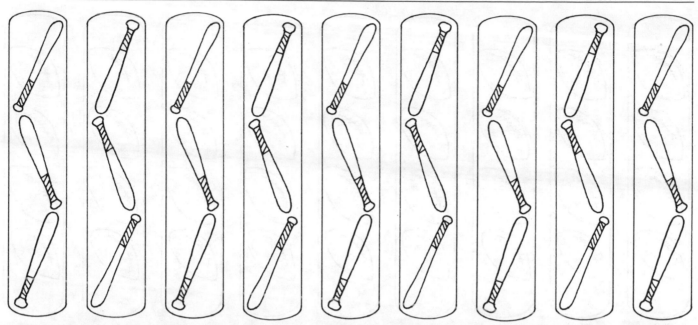

Multiply.

$1 \times 3 =$ _____ $4 \times 3 =$ _____ $7 \times 3 =$ _____

$2 \times 3 =$ _____ $5 \times 3 =$ _____ $8 \times 3 =$ _____

$3 \times 3 =$ _____ $6 \times 3 =$ _____ $9 \times 3 =$ _____

$$\begin{array}{r} 3 \\ \times 7 \\ \hline \end{array} \quad \begin{array}{r} 3 \\ \times 5 \\ \hline \end{array} \quad \begin{array}{r} 3 \\ \times 9 \\ \hline \end{array} \quad \begin{array}{r} 3 \\ \times 1 \\ \hline \end{array} \quad \begin{array}{r} 3 \\ \times 4 \\ \hline \end{array} \quad \begin{array}{r} 3 \\ \times 3 \\ \hline \end{array} \quad \begin{array}{r} 3 \\ \times 8 \\ \hline \end{array} \quad \begin{array}{r} 3 \\ \times 2 \\ \hline \end{array}$$

$$\begin{array}{r} 3 \\ \times 6 \\ \hline \end{array} \quad \begin{array}{r} 3 \\ \times 2 \\ \hline \end{array} \quad \begin{array}{r} 3 \\ \times 3 \\ \hline \end{array} \quad \begin{array}{r} 3 \\ \times 9 \\ \hline \end{array} \quad \begin{array}{r} 3 \\ \times 5 \\ \hline \end{array} \quad \begin{array}{r} 3 \\ \times 7 \\ \hline \end{array} \quad \begin{array}{r} 3 \\ \times 4 \\ \hline \end{array} \quad \begin{array}{r} 3 \\ \times 6 \\ \hline \end{array}$$

$$\begin{array}{r} 3 \\ \times 5 \\ \hline \end{array} \quad \begin{array}{r} 3 \\ \times 8 \\ \hline \end{array} \quad \begin{array}{r} 3 \\ \times 6 \\ \hline \end{array} \quad \begin{array}{r} 3 \\ \times 2 \\ \hline \end{array} \quad \begin{array}{r} 3 \\ \times 3 \\ \hline \end{array} \quad \begin{array}{r} 3 \\ \times 9 \\ \hline \end{array} \quad \begin{array}{r} 3 \\ \times 7 \\ \hline \end{array} \quad \begin{array}{r} 3 \\ \times 4 \\ \hline \end{array}$$

Name _____

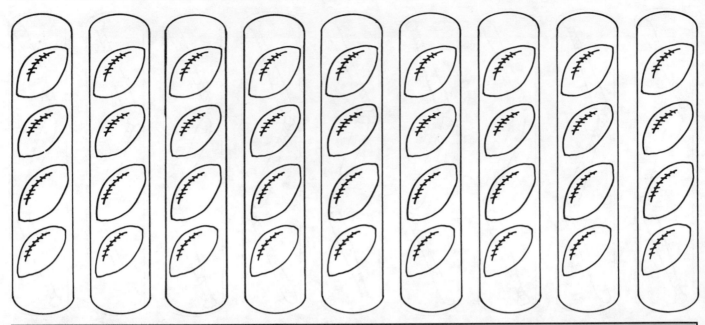

Multiply.

$1 \times 4 = $ _____	$4 \times 4 = $ _____	$7 \times 4 = $ _____
$2 \times 4 = $ _____	$5 \times 4 = $ _____	$8 \times 4 = $ _____
$3 \times 4 = $ _____	$6 \times 4 = $ _____	$9 \times 4 = $ _____

$$\begin{array}{cc} 4 \\ \times 6 \\ \hline \end{array} \quad \begin{array}{cc} 4 \\ \times 9 \\ \hline \end{array} \quad \begin{array}{cc} 4 \\ \times 4 \\ \hline \end{array} \quad \begin{array}{cc} 4 \\ \times 7 \\ \hline \end{array} \quad \begin{array}{cc} 4 \\ \times 5 \\ \hline \end{array} \quad \begin{array}{cc} 4 \\ \times 8 \\ \hline \end{array} \quad \begin{array}{cc} 4 \\ \times 3 \\ \hline \end{array} \quad \begin{array}{cc} 4 \\ \times 2 \\ \hline \end{array}$$

$$\begin{array}{cc} 4 \\ \times 7 \\ \hline \end{array} \quad \begin{array}{cc} 4 \\ \times 3 \\ \hline \end{array} \quad \begin{array}{cc} 4 \\ \times 6 \\ \hline \end{array} \quad \begin{array}{cc} 4 \\ \times 8 \\ \hline \end{array} \quad \begin{array}{cc} 4 \\ \times 2 \\ \hline \end{array} \quad \begin{array}{cc} 4 \\ \times 9 \\ \hline \end{array} \quad \begin{array}{cc} 4 \\ \times 5 \\ \hline \end{array} \quad \begin{array}{cc} 4 \\ \times 4 \\ \hline \end{array}$$

$$\begin{array}{cc} 4 \\ \times 9 \\ \hline \end{array} \quad \begin{array}{cc} 4 \\ \times 2 \\ \hline \end{array} \quad \begin{array}{cc} 4 \\ \times 7 \\ \hline \end{array} \quad \begin{array}{cc} 4 \\ \times 4 \\ \hline \end{array} \quad \begin{array}{cc} 4 \\ \times 5 \\ \hline \end{array} \quad \begin{array}{cc} 4 \\ \times 8 \\ \hline \end{array} \quad \begin{array}{cc} 4 \\ \times 1 \\ \hline \end{array} \quad \begin{array}{cc} 4 \\ \times 6 \\ \hline \end{array}$$

8

Multiplying 4

Name _____

$7 \times 2 =$ ____	$6 \times 5 =$ ____	$3 \times 3 =$ ____
$4 \times 4 =$ ____	$1 \times 4 =$ ____	$6 \times 2 =$ ____
$9 \times 3 =$ ____	$2 \times 2 =$ ____	$8 \times 4 =$ ____
$5 \times 5 =$ ____	$8 \times 2 =$ ____	$1 \times 5 =$ ____
$8 \times 5 =$ ____	$3 \times 2 =$ ____	$4 \times 3 =$ ____

$$\begin{array}{cccccccc} 2 & 3 & 5 & 3 & 5 & 3 & 3 & 4 \\ \times 9 & \times 8 & \times 7 & \times 9 & \times 6 & \times 7 & \times 4 & \times 5 \end{array}$$

$$\begin{array}{cccccccc} 5 & 2 & 4 & 5 & 5 & 4 & 5 & 4 \\ \times 8 & \times 4 & \times 6 & \times 3 & \times 9 & \times 7 & \times 2 & \times 9 \end{array}$$

$$\begin{array}{cccccccc} 2 & 4 & 2 & 3 & 4 & 3 & 5 & 2 \\ \times 3 & \times 3 & \times 1 & \times 5 & \times 2 & \times 6 & \times 4 & \times 5 \end{array}$$

Name _____

Alan works at a supermarket. Today he was busy restocking the shelves. Find out how many of each item he put on the shelves.

Solve.	
1. 7 rows of cake mix 4 boxes in each row	**2.** 6 rows of soap 2 bars in each row
3. 8 rows of flour 5 bags in each row	**4.** 7 rows of paper towels 3 rolls in each row
5. 5 rows of chicken soup 5 cans in each row	**6.** 8 rows of bread 2 loaves in each row
7. 3 rows of cat food 4 boxes in each row	**8.** 8 rows of cooking oil 4 bottles in each row
9. 6 rows of soft drink 5 bottles in each row	**10.** 9 rows of potato chips 5 packs in each row
11. 6 rows of peanut butter 4 jars in each row	**12.** 9 rows of tuna fish 2 cans in each row

Problem Solving
© The Continental Press, Inc.

Name _____

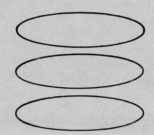

$3 \times 0 = 0$

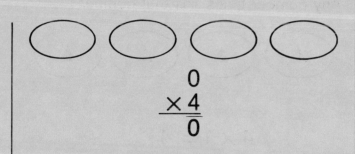

$$\begin{array}{r} 0 \\ \times\,4 \\ \hline 0 \end{array}$$

Multiply.

$1 \times 0 = \underline{0}$ $4 \times 0 = \underline{}$ $7 \times 0 = \underline{}$

$2 \times 0 = \underline{}$ $5 \times 0 = \underline{}$ $8 \times 0 = \underline{}$

$3 \times 0 = \underline{}$ $6 \times 0 = \underline{}$ $9 \times 0 = \underline{}$

| $\begin{array}{r}0\\\times6\\\hline 0\end{array}$ | $\begin{array}{r}3\\\times4\\\hline\end{array}$ | $\begin{array}{r}0\\\times9\\\hline\end{array}$ | $\begin{array}{r}4\\\times2\\\hline\end{array}$ | $\begin{array}{r}5\\\times1\\\hline\end{array}$ | $\begin{array}{r}0\\\times3\\\hline\end{array}$ | $\begin{array}{r}0\\\times4\\\hline\end{array}$ | $\begin{array}{r}5\\\times2\\\hline\end{array}$ |

| $\begin{array}{r}0\\\times5\\\hline\end{array}$ | $\begin{array}{r}4\\\times5\\\hline\end{array}$ | $\begin{array}{r}5\\\times9\\\hline\end{array}$ | $\begin{array}{r}0\\\times2\\\hline\end{array}$ | $\begin{array}{r}4\\\times4\\\hline\end{array}$ | $\begin{array}{r}0\\\times8\\\hline\end{array}$ | $\begin{array}{r}5\\\times6\\\hline\end{array}$ | $\begin{array}{r}3\\\times9\\\hline\end{array}$ |

| $\begin{array}{r}5\\\times7\\\hline\end{array}$ | $\begin{array}{r}3\\\times8\\\hline\end{array}$ | $\begin{array}{r}0\\\times7\\\hline\end{array}$ | $\begin{array}{r}4\\\times8\\\hline\end{array}$ | $\begin{array}{r}0\\\times1\\\hline\end{array}$ | $\begin{array}{r}2\\\times9\\\hline\end{array}$ | $\begin{array}{r}0\\\times4\\\hline\end{array}$ | $\begin{array}{r}4\\\times7\\\hline\end{array}$ |

Multiplying 0

Any number times 1 is that number.

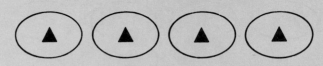

$4 \times 1 = 4$

$$\begin{array}{r} 1 \\ \times 2 \\ \hline 2 \end{array}$$

Multiply.

$1 \times 1 = \underline{1}$ $4 \times 1 = \underline{4}$ $7 \times 1 = \underline{7}$

$2 \times 1 = \underline{2}$ $5 \times 1 = \underline{5}$ $8 \times 1 = \underline{8}$

$3 \times 1 = \underline{3}$ $6 \times 1 = \underline{6}$ $9 \times 1 = \underline{9}$

$$\begin{array}{r} 1 \\ \times 5 \\ \hline 5 \end{array} \qquad \begin{array}{r} 2 \\ \times 4 \\ \hline 8 \end{array} \qquad \begin{array}{r} 3 \\ \times 5 \\ \hline 15 \end{array} \qquad \begin{array}{r} 1 \\ \times 9 \\ \hline 9 \end{array} \qquad \begin{array}{r} 0 \\ \times 7 \\ \hline 0 \end{array} \qquad \begin{array}{r} 4 \\ \times 6 \\ \hline 24 \end{array}$$

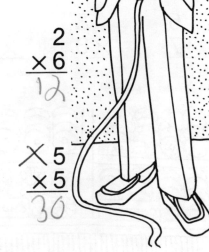

$$\begin{array}{r} 1 \\ \times 3 \\ \hline 3 \end{array} \qquad \begin{array}{r} 5 \\ \times 8 \\ \hline 40 \end{array} \qquad \begin{array}{r} 1 \\ \times 7 \\ \hline 7 \end{array} \qquad \begin{array}{r} 3 \\ \times 3 \\ \hline 9 \end{array} \qquad \begin{array}{r} 2 \\ \times 8 \\ \hline 16 \end{array} \qquad \begin{array}{r} 1 \\ \times 6 \\ \hline 6 \end{array}$$

$$\begin{array}{r} 5 \\ \times 4 \\ \hline 25 \end{array} \qquad \begin{array}{r} 1 \\ \times 0 \\ \hline 0 \end{array} \qquad \begin{array}{r} 1 \\ \times 4 \\ \hline 4 \end{array} \qquad \begin{array}{r} 1 \\ \times 2 \\ \hline 2 \end{array} \qquad \begin{array}{r} 3 \\ \times 8 \\ \hline 24 \end{array} \qquad \begin{array}{r} 2 \\ \times 6 \\ \hline 12 \end{array}$$

$$\begin{array}{r} 4 \\ \times 3 \\ \hline 12 \end{array} \qquad \begin{array}{r} 1 \\ \times 1 \\ \hline 1 \end{array} \qquad \begin{array}{r} 1 \\ \times 8 \\ \hline 8 \end{array} \qquad \begin{array}{r} 2 \\ \times 3 \\ \hline 6 \end{array} \qquad \begin{array}{r} 4 \\ \times 9 \\ \hline 32 \end{array} \qquad \begin{array}{r} 5 \\ \times 5 \\ \hline 30 \end{array}$$

Multiplying 1
© The Continental Press, Inc.

Math

Name Maliah West

 96% ☺ 2/10/05

Multiply.

Very Good!

$4 \times 0 = 0$ ✓

$5 \times 3 = 15$ ✓

$6 \times 5 = 30$ ✓

$2 \times 1 = 2$ ✓

$3 \times 2 = 6$ ✓

$9 \times 4 = 36$ ✓

$5 \times 5 = 25$ ✓

$9 \times 1 = 9$ ✓

$8 \times 2 = 16$ ✓

$4 \times 5 = 20$ ✓

$1 \times 5 = 5$ ✓

$8 \times 0 = 0$ ✓

$6 \times 1 = 6$ ✓

$9 \times 3 = 27$ ✓

$5 \times 2 = 10$ ✓

$3 \times 3 = 9$ ✓

$7 \times 4 = 28$ ✓

$1 \times 1 = 1$ ✓

$9 \times 2 = 18$ ✓

$6 \times 3 = 18$ ✓

$7 \times 5 = 25$ ✗

$$\begin{array}{r} 5 \\ \times 9 \\ \hline 45 \end{array}$$ ✓

$$\begin{array}{r} 0 \\ \times 5 \\ \hline 0 \end{array}$$ ✓

$$\begin{array}{r} 3 \\ \times 8 \\ \hline 24 \end{array}$$ ✓

$$\begin{array}{r} 4 \\ \times 2 \\ \hline 8 \end{array}$$ ✓

$$\begin{array}{r} 3 \\ \times 9 \\ \hline 27 \end{array}$$ ✓

$$\begin{array}{r} 4 \\ \times 4 \\ \hline 16 \end{array}$$ ✓

$$\begin{array}{r} 5 \\ \times 8 \\ \hline 40 \end{array}$$ ✓

$$\begin{array}{r} 0 \\ \times 6 \\ \hline 0 \end{array}$$ ✓

$$\begin{array}{r} 3 \\ \times 1 \\ \hline 3 \end{array}$$ ✓

$$\begin{array}{r} 4 \\ \times 6 \\ \hline 24 \end{array}$$ ✓

$$\begin{array}{r} 3 \\ \times 7 \\ \hline 21 \end{array}$$ ✓

$$\begin{array}{r} 2 \\ \times 6 \\ \hline 12 \end{array}$$ ✓

$$\begin{array}{r} 1 \\ \times 4 \\ \hline 4 \end{array}$$ ✓

$$\begin{array}{r} 4 \\ \times 5 \\ \hline 20 \end{array}$$ ✓

$$\begin{array}{r} 4 \\ \times 8 \\ \hline 32 \end{array}$$ ✓

Review: Multiplying 0–5
© The Continental Press, Inc.

13

Name _____

1. Karen made 8 batches of peanut candy for gifts. She used 1 cup of peanuts in each batch. How many cups of peanuts did she use altogether?

   8   8×1=8
   8 peanuts

2. Sam jogs 2 miles every day after school. How far does he jog in 7 days?

   7 × 2 = 14 miles

3. Jenny writes 3 letters to her pen pals each week. How many letters does she write in 5 weeks?

4. At the car wash, Ted cleans 4 cars an hour. How many would he clean in 4 hours?

5. Marcia has a part-time job. She works 3 hours a day. How many hours would she work in 7 days?

6. Lee bought 8 pounds of ground meat for a cookout. If he can make 4 burgers from a pound of meat, how many burgers can he make in all?

7. Ms. Trader gets about 5 messages a day on her telephone answering machine. How many messages would she get in 5 days?

8. Kathy can swim 2 laps in a minute. How many laps can she swim in 8 minutes?

14

Problem Solving
© The Continental Press, Inc.

Name _____

2/3/05

Multiply.

$1 \times 6 = 6$

$2 \times 6 = 12$

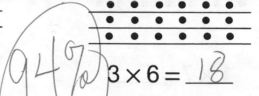

94%

$3 \times 6 = 18$

$4 \times 6 = 24$

$5 \times 6 = 30$

$6 \times 6 = 36$

$7 \times 6 = 42$

$8 \times 6 = 48$

$9 \times 6 = 53$

$$\begin{array}{c} 6 \\ \times 3 \\ \hline 18 \end{array} \quad \begin{array}{c} 6 \\ \times 1 \\ \hline 6 \end{array} \quad \begin{array}{c} 6 \\ \times 5 \\ \hline 30 \end{array} \quad \begin{array}{c} 6 \\ \times 8 \\ \hline 48 \end{array} \quad \begin{array}{c} 6 \\ \times 6 \\ \hline 36 \end{array} \quad \begin{array}{c} 6 \\ \times 9 \\ \hline 53 \end{array} \quad \begin{array}{c} 6 \\ \times 2 \\ \hline 18 \end{array} \quad \begin{array}{c} 6 \\ \times 7 \\ \hline 42 \end{array}$$

$$\begin{array}{c} 6 \\ \times 1 \\ \hline 6 \end{array} \quad \begin{array}{c} 6 \\ \times 3 \\ \hline 18 \end{array} \quad \begin{array}{c} 6 \\ \times 7 \\ \hline 42 \end{array} \quad \begin{array}{c} 6 \\ \times 9 \\ \hline 53 \end{array} \quad \begin{array}{c} 6 \\ \times 5 \\ \hline 30 \end{array} \quad \begin{array}{c} 6 \\ \times 4 \\ \hline 24 \end{array} \quad \begin{array}{c} 6 \\ \times 6 \\ \hline 36 \end{array} \quad \begin{array}{c} 6 \\ \times 2 \\ \hline 18 \end{array}$$

$$\begin{array}{c} 6 \\ \times 7 \\ \hline 42 \end{array} \quad \begin{array}{c} 6 \\ \times 8 \\ \hline 48 \end{array} \quad \begin{array}{c} 6 \\ \times 2 \\ \hline 12 \end{array} \quad \begin{array}{c} 6 \\ \times 6 \\ \hline 36 \end{array} \quad \begin{array}{c} 6 \\ \times 9 \\ \hline 53 \end{array} \quad \begin{array}{c} 6 \\ \times 8 \\ \hline 48 \end{array} \quad \begin{array}{c} 6 \\ \times 3 \\ \hline 18 \end{array} \quad \begin{array}{c} 6 \\ \times 4 \\ \hline 24 \end{array}$$

Multiplying 6
© The Continental Press, Inc.

15

Name _____

· · · · · · ·

$1 \times 7 = \underline{7}$ ✓

· · · · · · ·
· · · · · · ·

$2 \times 7 = \underline{14}$ ✓

· · · · · · ·
· · · · · · ·
· · · · · · ·

$3 \times 7 = \underline{21}$ ✓

· · · · · · ·
· · · · · · ·
· · · · · · ·
· · · · · · ·

$4 \times 7 = \underline{28}$ ✓

· · · · · · ·
· · · · · · ·
· · · · · · ·
· · · · · · ·
· · · · · · ·

$5 \times 7 = \underline{35}$ ✓

· · · · · · ·
· · · · · · ·
· · · · · · ·
· · · · · · ·
· · · · · · ·
· · · · · · ·

$6 \times 7 = \underline{42}$ ✓

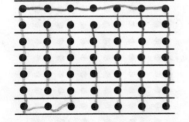

$7 \times 7 = \underline{49}$ ✓

· · · · · · ·
· · · · · · ·
· · · · · · ·
· · · · · · ·
· · · · · · ·
· · · · · · ·
· · · · · · ·
· · · · · · ·

$8 \times 7 = \underline{56}$ ✓

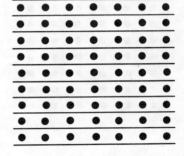

$9 \times 7 = \underline{63}$ ✓

$\begin{array}{r}7\\ \times 4\\ \hline\end{array}$ ✓	$\begin{array}{r}7\\ \times 2\\ \hline\end{array}$ ✓	$\begin{array}{r}7\\ \times 7\\ \hline\end{array}$ ✓	$\begin{array}{r}7\\ \times 5\\ \hline\end{array}$ ✓	$\begin{array}{r}7\\ \times 9\\ \hline\end{array}$ ✓	$\begin{array}{r}7\\ \times 1\\ \hline\end{array}$ ✓	$\begin{array}{r}7\\ \times 3\\ \hline\end{array}$ ✓	$\begin{array}{r}7\\ \times 8\\ \hline\end{array}$ ✓
28	14	49	35	63	7	21	56

$\begin{array}{r}7\\ \times 5\\ \hline\end{array}$ ✓	$\begin{array}{r}7\\ \times 7\\ \hline\end{array}$ ✓	$\begin{array}{r}7\\ \times 6\\ \hline\end{array}$ ✓	$\begin{array}{r}7\\ \times 3\\ \hline\end{array}$ ✓	$\begin{array}{r}7\\ \times 2\\ \hline\end{array}$ ✓	$\begin{array}{r}7\\ \times 9\\ \hline\end{array}$ ✓	$\begin{array}{r}7\\ \times 8\\ \hline\end{array}$ ✓	$\begin{array}{r}7\\ \times 4\\ \hline\end{array}$
35	49	42	21	14	63	56	28

$\begin{array}{r}7\\ \times 6\\ \hline\end{array}$ ✓	$\begin{array}{r}7\\ \times 1\\ \hline\end{array}$ ✓	$\begin{array}{r}7\\ \times 5\\ \hline\end{array}$ ✓	$\begin{array}{r}7\\ \times 8\\ \hline\end{array}$ ✓	$\begin{array}{r}7\\ \times 3\\ \hline\end{array}$ ✓	$\begin{array}{r}7\\ \times 4\\ \hline\end{array}$ ✓	$\begin{array}{r}7\\ \times 2\\ \hline\end{array}$ ✓	$\begin{array}{r}7\\ \times 9\\ \hline\end{array}$ ✓
42	7	35	56	21	28	14	63

Name _____

$1 \times 8 = 8$	
$2 \times 8 = 16$	
$3 \times 8 = 24$	
$4 \times 8 = 32$	
$5 \times 8 = 40$	
$6 \times 8 = 48$	
$7 \times 8 = 56$	
$8 \times 8 = 64$	
$9 \times 8 = 72$	

Multiply.

$4 \times 8 = $ ___

$8 \times 8 = $ ___

$1 \times 8 = $ ___

$7 \times 8 = $ ___

$2 \times 8 = $ ___

$6 \times 8 = $ ___

$5 \times 8 = $ ___

$9 \times 8 = $ ___

$3 \times 8 = $ ___

Multiply.

8 ×3	8 ×1	8 ×6	8 ×9	8 ×7	8 ×4	8 ×2	8 ×5
8 ×8	8 ×4	8 ×7	8 ×2	8 ×9	8 ×3	8 ×6	8 ×8
8 ×5	8 ×9	8 ×6	8 ×8	8 ×4	8 ×2	8 ×3	8 ×7

Name _____

Study these multiplication facts for 9.

$1 \times 9 = 9$

$2 \times 9 = 18$

$3 \times 9 = 27$

$4 \times 9 = 36$

$5 \times 9 = 45$

$6 \times 9 = 54$

$7 \times 9 = 63$

$8 \times 9 = 72$

$9 \times 9 = 81$

Multiply.

$5 \times 9 = $ _____

$3 \times 9 = $ _____

$1 \times 9 = $ _____

$9 \times 9 = $ _____

$7 \times 9 = $ _____

$2 \times 9 = $ _____

$6 \times 9 = $ _____

$4 \times 9 = $ _____

$8 \times 9 = $ _____

Multiply.

$$\begin{array}{cc} 9 \\ \times 4 \end{array} \quad \begin{array}{cc} 9 \\ \times 2 \end{array} \quad \begin{array}{cc} 9 \\ \times 6 \end{array} \quad \begin{array}{cc} 9 \\ \times 8 \end{array} \quad \begin{array}{cc} 9 \\ \times 1 \end{array} \quad \begin{array}{cc} 9 \\ \times 3 \end{array} \quad \begin{array}{cc} 9 \\ \times 5 \end{array} \quad \begin{array}{cc} 9 \\ \times 7 \end{array}$$

$$\begin{array}{cc} 9 \\ \times 3 \end{array} \quad \begin{array}{cc} 9 \\ \times 6 \end{array} \quad \begin{array}{cc} 9 \\ \times 7 \end{array} \quad \begin{array}{cc} 9 \\ \times 9 \end{array} \quad \begin{array}{cc} 9 \\ \times 2 \end{array} \quad \begin{array}{cc} 9 \\ \times 4 \end{array} \quad \begin{array}{cc} 9 \\ \times 8 \end{array} \quad \begin{array}{cc} 9 \\ \times 5 \end{array}$$

$$\begin{array}{cc} 9 \\ \times 7 \end{array} \quad \begin{array}{cc} 9 \\ \times 9 \end{array} \quad \begin{array}{cc} 9 \\ \times 1 \end{array} \quad \begin{array}{cc} 9 \\ \times 8 \end{array} \quad \begin{array}{cc} 9 \\ \times 4 \end{array} \quad \begin{array}{cc} 9 \\ \times 2 \end{array} \quad \begin{array}{cc} 9 \\ \times 3 \end{array} \quad \begin{array}{cc} 9 \\ \times 6 \end{array}$$

18

Multiplying 9

Multiply.

$3 \times 8 =$ ___	$3 \times 7 =$ ___	$2 \times 6 =$ ___
$6 \times 9 =$ ___	$6 \times 6 =$ ___	$7 \times 7 =$ ___
$8 \times 7 =$ ___	$1 \times 8 =$ ___	$5 \times 8 =$ ___
$5 \times 6 =$ ___	$9 \times 7 =$ ___	$9 \times 9 =$ ___
$4 \times 8 =$ ___	$3 \times 9 =$ ___	$4 \times 6 =$ ___
$1 \times 6 =$ ___	$6 \times 8 =$ ___	$6 \times 7 =$ ___
$8 \times 9 =$ ___	$5 \times 7 =$ ___	$7 \times 8 =$ ___

$$\begin{array}{cccccccc} 6 & 7 & 9 & 7 & 7 & 7 & 9 & 8 \\ \times 3 & \times 4 & \times 1 & \times 9 & \times 2 & \times 8 & \times 5 & \times 8 \end{array}$$

$$\begin{array}{cccccccc} 8 & 7 & 6 & 9 & 8 & 9 & 8 & 6 \\ \times 2 & \times 1 & \times 7 & \times 2 & \times 6 & \times 4 & \times 9 & \times 9 \end{array}$$

Name _____

Solve.

1. The school band meets 9 times a month for 9 months of the year. How many times does it meet in a year?

2. There are 9 rows in the band. The band leader put 7 students in each row. How many students are in the band?

3. There are 7 students in each row of the band. If 4 rows are horn players, how many students play horns?

4. Last week the band practiced a special tune 4 times. If the tune takes 6 minutes to play, how long did the band spend practicing it?

5. The band played 5 songs at the spring concert. Each song was 8 minutes long. How long did the band play?

6. At each football game the band plays 6 tunes. How many tunes will it play at 8 football games?

7. The program has 3 parts. If the band marches 9 blocks during each part, how far will it have marched to do the entire program?

8. A special marching drill takes 8 minutes. The band did it 4 times in a parade. How much time did the band spend on it?

Name _____

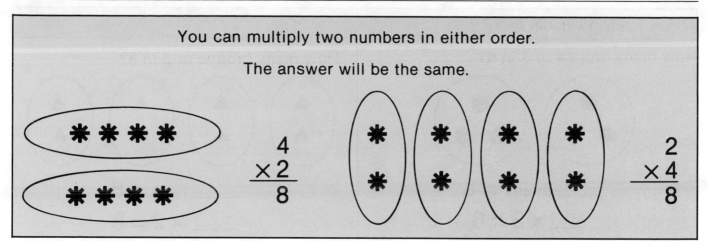

Multiply.

$5 \times 2 = \underline{10}$ $9 \times 4 = \underline{}$ $1 \times 7 = \underline{}$

$2 \times 5 = \underline{}$ $4 \times 9 = \underline{}$ $7 \times 1 = \underline{}$

$3 \times 6 = \underline{}$ $0 \times 3 = \underline{}$ $5 \times 8 = \underline{}$

$6 \times 3 = \underline{}$ $3 \times 0 = \underline{}$ $8 \times 5 = \underline{}$

$$\begin{array}{cccccccc} 8 & 0 & 7 & 2 & 4 & 5 & 1 & 6 \\ \times 0 & \times 8 & \times 2 & \times 7 & \times 5 & \times 4 & \times 6 & \times 1 \\ \hline 0 & & & & & & & \end{array}$$

$$\begin{array}{cccccccc} 9 & 8 & 1 & 6 & 0 & 4 & 6 & 8 \\ \times 8 & \times 9 & \times 6 & \times 1 & \times 4 & \times 0 & \times 8 & \times 6 \\ \hline \end{array}$$

$$\begin{array}{cccccccc} 3 & 2 & 9 & 5 & 7 & 6 & 4 & 8 \\ \times 2 & \times 3 & \times 5 & \times 9 & \times 6 & \times 7 & \times 8 & \times 4 \\ \hline \end{array}$$

Name _____

How many groups in each?

How many groups of 3 in 6?

_____ threes = 6

_____ × 3 = 6

How many groups of 2 in 8?

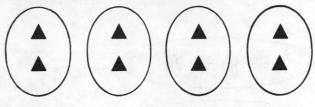

_____ twos = 8

_____ × 2 = 8

How many groups of 5 in 10?

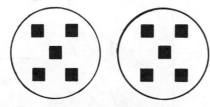

_____ fives = 10

_____ × 5 = 10

How many groups of 4 in 12?

_____ fours = 12

_____ × 4 = 12

When you find a missing factor, you are dividing. The sign ÷ means divided by.

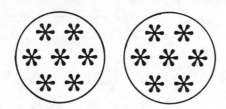

_____ × 7 = 14

14 divided by 7 = _____

14 ÷ 7 = _____

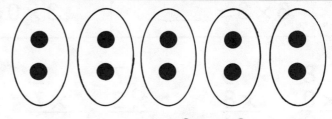

_____ × 2 = 10

10 divided by 2 = _____

10 ÷ 2 = _____

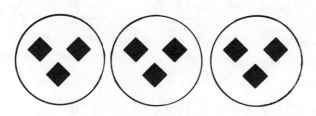

_____ × 3 = 9

9 divided by 3 = _____

9 ÷ 3 = _____

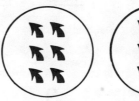

_____ × 6 = 12

12 divided by 6 = _____

12 ÷ 6 = _____

22

Division as Related to Multiplication
© The Continental Press, Inc.

Name _____

Divide to find a missing factor.

$$5 \times 3 = 15 \qquad 15 \div 3 = 5$$

product

factor

factor

multiplication sentence   division sentence

missing factor → $5 \times 3 = 15$ $15 \div 3 = 5$ ← quotient

Divide.

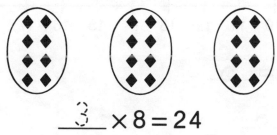

___3___ $\times 8 = 24$

24 divided by $8 =$ ___3___

$24 \div 8 =$ ___3___

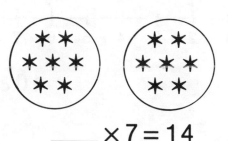

_____ $\times 7 = 14$

14 divided by $7 =$ _____

$14 \div 7 =$ _____

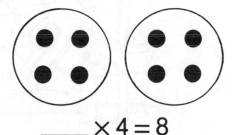

_____ $\times 4 = 8$

8 divided by $4 =$ _____

$8 \div 4 =$ _____

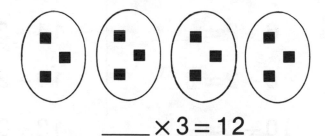

_____ $\times 3 = 12$

12 divided by $3 =$ _____

$12 \div 3 =$ _____

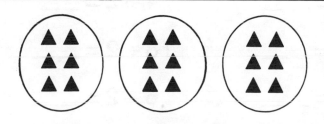

_____ $\times 6 = 18$

18 divided by $6 =$ _____

$18 \div 6 =$ _____

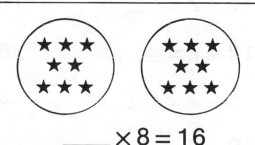

_____ $\times 8 = 16$

16 divided by $8 =$ _____

$16 \div 8 =$ _____

Introduction to Division
© The Continental Press, Inc.

23

Name _____

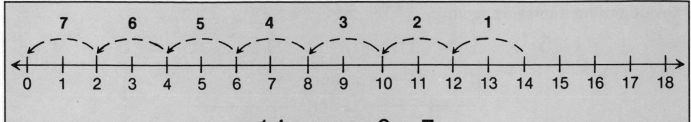

14 divided by 2 = 7

14 ÷ 2 = 7

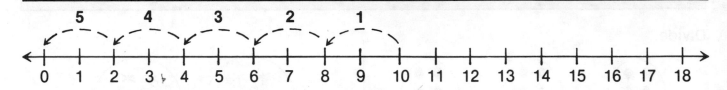

10 divided by 2 = __5__

10 ÷ 2 = __5__

Use the number line to help you divide.

4 ÷ 2 = __2__   12 ÷ 2 = __6__

6 ÷ 2 = __3__   14 ÷ 2 = __7__

8 ÷ 2 = __4__   16 ÷ 2 = __8__

10 ÷ 2 = __5__   18 ÷ 2 = __9__

Divide.

16 ÷ 2 = __8__   10 ÷ 2 = __5__   4 ÷ 2 = __2__

12 ÷ 2 = __6__   8 ÷ 2 = __4__   18 ÷ 2 = __9__

6 ÷ 2 = __3__   4 ÷ 2 = __2__   6 ÷ 2 = __3__

10 ÷ 2 = __5__   12 ÷ 2 = __6__   14 ÷ 2 = __7__

14 ÷ 2 = __7__   18 ÷ 2 = __9__   16 ÷ 2 = __8__

24

Dividing by 2

Name _____

A division problem can be written two different ways.

Number line from 0 to 50 marked in intervals of 5, with dashed arrows labeled 7, 6, 5, 4, 3, 2, 1.

$$35 \div 5 = 7 \qquad 5{\overline{\smash{\big)}\,35}}$$

dividend divisor quotient   divisor

7 ← quotient

35 ← dividend

Use the number line to help you divide.

$10 \div 5 = \underline{2}$ $30 \div 5 = \underline{6}$

$15 \div 5 = \underline{3}$ $35 \div 5 = \underline{7}$

$20 \div 5 = \underline{4}$ $40 \div 5 = \underline{8}$

$25 \div 5 = \underline{5}$ $45 \div 5 = \underline{9}$

Divide.

$5{\overline{\smash{\big)}\,15}} = 3 \qquad 5{\overline{\smash{\big)}\,25}} = 5 \qquad 5{\overline{\smash{\big)}\,30}} = 6 \qquad 5{\overline{\smash{\big)}\,10}} = 2 \qquad 5{\overline{\smash{\big)}\,40}} = 8$

$5{\overline{\smash{\big)}\,35}} = 7 \qquad 5{\overline{\smash{\big)}\,10}} = 2 \qquad 5{\overline{\smash{\big)}\,45}} = 4 \qquad 5{\overline{\smash{\big)}\,30}} = 6 \qquad 5{\overline{\smash{\big)}\,25}} = 4$

$5{\overline{\smash{\big)}\,20}} = 4 \qquad 5{\overline{\smash{\big)}\,40}} = 8 \qquad 5{\overline{\smash{\big)}\,35}} = 7 \qquad 5{\overline{\smash{\big)}\,25}} = 5 \qquad 5{\overline{\smash{\big)}\,30}} = 6$

$5{\overline{\smash{\big)}\,45}} = 9 \qquad 5{\overline{\smash{\big)}\,15}} = 3 \qquad 5{\overline{\smash{\big)}\,20}} = 4 \qquad 5{\overline{\smash{\big)}\,40}} = 8 \qquad 5{\overline{\smash{\big)}\,10}} = 2$

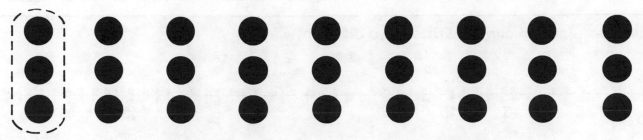

How many groups of 3 are in 27? ___9___ Circle them.

$$27 \div 3 = \underline{9} \qquad 3\overline{)27}^{\,9}$$

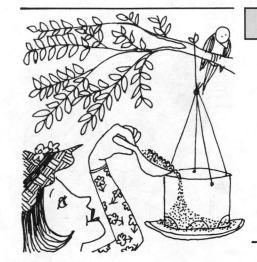

Divide.	
$6 \div 3 = \underline{}$	$18 \div 3 = \underline{}$
$9 \div 3 = \underline{}$	$21 \div 3 = \underline{}$
$12 \div 3 = \underline{}$	$24 \div 3 = \underline{}$
$15 \div 3 = \underline{}$	$27 \div 3 = \underline{}$

$3\overline{)27} \qquad 3\overline{)18} \qquad 3\overline{)24} \qquad 3\overline{)9} \qquad 3\overline{)15}$

$3\overline{)18} \qquad 3\overline{)27} \qquad 3\overline{)12} \qquad 3\overline{)6} \qquad 3\overline{)21}$

$3\overline{)12} \qquad 3\overline{)15} \qquad 3\overline{)27} \qquad 3\overline{)24} \qquad 3\overline{)6}$

$3\overline{)24} \qquad 3\overline{)9} \qquad 3\overline{)21} \qquad 3\overline{)12} \qquad 3\overline{)18}$

Dividing by 3

Name _____

How many groups of 4 are in 36? ____9____ Circle them.

$$36 \div 4 = \underline{9} \qquad 4\overline{)36}$$

Divide.		
$8 \div 4 = \underline{\quad}$	$24 \div 4 = \underline{\quad}$	
$12 \div 4 = \underline{\quad}$	$28 \div 4 = \underline{\quad}$	
$16 \div 4 = \underline{\quad}$	$32 \div 4 = \underline{\quad}$	
$20 \div 4 = \underline{\quad}$	$36 \div 4 = \underline{\quad}$	

$4\overline{)20} \qquad 4\overline{)8} \qquad 4\overline{)12} \qquad 4\overline{)28} \qquad 4\overline{)32}$

$4\overline{)28} \qquad 4\overline{)36} \qquad 4\overline{)16} \qquad 4\overline{)24} \qquad 4\overline{)20}$

$4\overline{)12} \qquad 4\overline{)32} \qquad 4\overline{)20} \qquad 4\overline{)36} \qquad 4\overline{)8}$

$4\overline{)16} \qquad 4\overline{)24} \qquad 4\overline{)12} \qquad 4\overline{)32} \qquad 4\overline{)28}$

Name _____

$12 \div 4 = $ ____	$30 \div 5 = $ ____	$14 \div 2 = $ ____
$15 \div 3 = $ ____	$20 \div 4 = $ ____	$9 \div 3 = $ ____
$16 \div 2 = $ ____	$18 \div 3 = $ ____	$35 \div 5 = $ ____
$27 \div 3 = $ ____	$45 \div 5 = $ ____	$8 \div 2 = $ ____

$4\overline{)36}$ $2\overline{)12}$ $5\overline{)25}$ $3\overline{)21}$

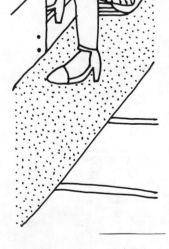

$2\overline{)18}$ $4\overline{)28}$ $2\overline{)6}$ $5\overline{)10}$

$3\overline{)24}$ $4\overline{)8}$ $3\overline{)6}$ $4\overline{)16}$

$5\overline{)20}$ $2\overline{)10}$ $4\overline{)32}$ $5\overline{)15}$

$3\overline{)12}$ $4\overline{)24}$ $5\overline{)40}$ $2\overline{)4}$

28

Name _____

Any number divided by 1 is that number.

How many groups of 1 in 4?   4

$4 \div 1 = 4$ $1\overline{)4}$ (quotient 4)

How many groups of 1 in 6? _6_

$6 \div 1 = 6$ $1\overline{)6}$ (quotient 6)

Divide.

$1 \div 1 =$ ____ $4 \div 1 =$ ____ $7 \div 1 =$ ____

$2 \div 1 =$ ____ $5 \div 1 =$ ____ $8 \div 1 =$ ____

$3 \div 1 =$ ____ $6 \div 1 =$ ____ $9 \div 1 =$ ____

$1\overline{)7}$ $1\overline{)9}$ $1\overline{)5}$ $1\overline{)2}$ $1\overline{)8}$ $1\overline{)3}$

Any number divided by itself is 1.

How many groups of 5 in 5?   1

$5 \div 5 = 1$ $5\overline{)5}$ (quotient 1)

How many groups of 7 in 7? _1_

$7 \div 7 = 1$ $7\overline{)7}$ (quotient 1)

Divide.

$1 \div 1 =$ ____ $4 \div 4 =$ ____ $7 \div 7 =$ ____

$2 \div 2 =$ ____ $5 \div 5 =$ ____ $8 \div 8 =$ ____

$3 \div 3 =$ ____ $6 \div 6 =$ ____ $9 \div 9 =$ ____

$2\overline{)2}$ $8\overline{)8}$ $6\overline{)6}$ $3\overline{)3}$ $9\overline{)9}$ $4\overline{)4}$

Dividing by 1 and Dividing a Number by Itself

Name _____

Solve.

1. The Romitos moved to a new house. It has 9 rooms. If there are 3 rooms on each floor, how many floors does the house have?

2. Barb sent 45 postcards to tell people she had moved. If she wrote 5 postcards each day, how many days did it take her to write all the postcards?

3. Mr. Danner typed 20 letters in 5 hours. How many letters did he type in an hour?

4. A dancing teacher had 18 people in her class. She divided the class into groups of 2. How many pairs of people were in the class?

5. Adam paid $24 for 3 tickets to a rock concert. How much did each ticket cost?

6. A waitress set 32 places in a restaurant. If she set 4 places at each table, how many tables did she set?

7. A steep trail up a mountain is 5 miles long. If Dan covers 1 mile in an hour, how long will it take him to hike the whole trail?

8. Abby spent $7 on 7 books. How much did each book cost?

Problem Solving
© The Continental Press, Inc.

Name _____

• • • • • •

$6 \div 6 =$ _____

• • • • • •
• • • • • •

$12 \div 6 =$ _____

• • • • • •
• • • • • •
• • • • • •

$18 \div 6 =$ _____

• • • • • •
• • • • • •
• • • • • •
• • • • • •

$24 \div 6 =$ _____

• • • • • •
• • • • • •
• • • • • •
• • • • • •
• • • • • •

$30 \div 6 =$ _____

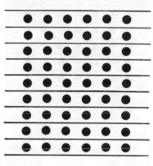

$36 \div 6 =$ _____

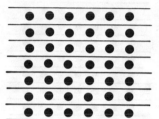

$42 \div 6 =$ _____

• • • • • •
• • • • • •
• • • • • •
• • • • • •
• • • • • •
• • • • • •
• • • • • •
• • • • • •

$48 \div 6 =$ _____

• • • • • •
• • • • • •
• • • • • •
• • • • • •
• • • • • •
• • • • • •
• • • • • •
• • • • • •
• • • • • •

$54 \div 6 =$ _____

$6 \overline{)48}$ $6 \overline{)6}$ $6 \overline{)12}$ $6 \overline{)30}$ $6 \overline{)54}$

$6 \overline{)18}$ $6 \overline{)24}$ $6 \overline{)42}$ $6 \overline{)36}$ $6 \overline{)12}$

$6 \overline{)24}$ $6 \overline{)18}$ $6 \overline{)6}$ $6 \overline{)48}$ $6 \overline{)36}$

$6 \overline{)30}$ $6 \overline{)42}$ $6 \overline{)24}$ $6 \overline{)54}$ $6 \overline{)18}$

Name _____

• • • • • • •

$7 \div 7 = $ _____

• • • • • • •
• • • • • • •

$14 \div 7 = $ _____

• • • • • • •
• • • • • • •
• • • • • • •

$21 \div 7 = $ _____

• • • • • • •
• • • • • • •
• • • • • • •
• • • • • • •

$28 \div 7 = $ _____

• • • • • • •
• • • • • • •
• • • • • • •
• • • • • • •
• • • • • • •

$35 \div 7 = $ _____

• • • • • • •
• • • • • • •
• • • • • • •
• • • • • • •
• • • • • • •
• • • • • • •

$42 \div 7 = $ _____

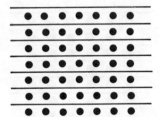

$49 \div 7 = $ _____

$56 \div 7 = $ _____

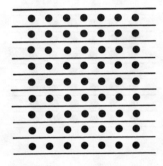

$63 \div 7 = $ _____

$7 \overline{)42}$ $7 \overline{)63}$ $7 \overline{)14}$ $7 \overline{)7}$ $7 \overline{)49}$

$7 \overline{)28}$ $7 \overline{)35}$ $7 \overline{)21}$ $7 \overline{)56}$ $7 \overline{)14}$

$7 \overline{)21}$ $7 \overline{)56}$ $7 \overline{)35}$ $7 \overline{)49}$ $7 \overline{)63}$

$7 \overline{)7}$ $7 \overline{)42}$ $7 \overline{)14}$ $7 \overline{)28}$ $7 \overline{)56}$

Name _____

Study these division facts for 8.

$$8 \div 8 = 1$$

$$16 \div 8 = 2$$

$$24 \div 8 = 3$$

$$32 \div 8 = 4$$

$$40 \div 8 = 5$$

$$48 \div 8 = 6$$

$$56 \div 8 = 7$$

$$64 \div 8 = 8$$

$$72 \div 8 = 9$$

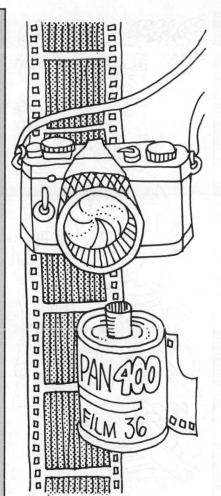

Divide.

$$32 \div 8 = \underline{\quad}$$

$$16 \div 8 = \underline{\quad}$$

$$48 \div 8 = \underline{\quad}$$

$$56 \div 8 = \underline{\quad}$$

$$8 \div 8 = \underline{\quad}$$

$$40 \div 8 = \underline{\quad}$$

$$72 \div 8 = \underline{\quad}$$

$$64 \div 8 = \underline{\quad}$$

$$24 \div 8 = \underline{\quad}$$

Divide.

$$8 \overline{)32} \qquad 8 \overline{)72} \qquad 8 \overline{)40} \qquad 8 \overline{)16} \qquad 8 \overline{)48}$$

$$8 \overline{)8} \qquad 8 \overline{)24} \qquad 8 \overline{)56} \qquad 8 \overline{)72} \qquad 8 \overline{)64}$$

$$8 \overline{)40} \qquad 8 \overline{)32} \qquad 8 \overline{)16} \qquad 8 \overline{)48} \qquad 8 \overline{)56}$$

$$8 \overline{)72} \qquad 8 \overline{)8} \qquad 8 \overline{)64} \qquad 8 \overline{)56} \qquad 8 \overline{)24}$$

Dividing by 8

Name _____

<table>
<tr><td>Study these division facts for 9.</td></tr>
</table>

$9 \div 9 = 1$

$18 \div 9 = 2$

$27 \div 9 = 3$

$36 \div 9 = 4$

$45 \div 9 = 5$

$54 \div 9 = 6$

$63 \div 9 = 7$

$72 \div 9 = 8$

$81 \div 9 = 9$

Divide.

$45 \div 9 = \underline{6}$ ✗

$63 \div 9 = \underline{7}$ ✓

$27 \div 9 = \underline{3}$ ✓

$81 \div 9 = \underline{9}$ ✓

$54 \div 9 = \underline{6}$ ✓

$9 \div 9 = \underline{81}$ ✗

$72 \div 9 = \underline{8}$ ✓

$18 \div 9 = \underline{2}$ ✓

$36 \div 9 = \underline{4}$ ✓

Divide.

$9\overline{)36}$ $\overset{4}{}$ ✓ $9\overline{)54}$ $\overset{6}{}$ ✓ $9\overline{)9}$ $\overset{81}{}$ ✗ $9\overline{)81}$ $\overset{9}{}$ $9\overline{)18}$ $\overset{3}{}$ ✓

$9\overline{)27}$ $\overset{3}{}$ ✓ $9\overline{)45}$ $\overset{5}{}$ ✓ $9\overline{)63}$ $\overset{7}{}$ $9\overline{)36}$ $\overset{4}{}$ $9\overline{)72}$ $\overset{8}{}$

$9\overline{)54}$ $\overset{6}{}$ ✓ $9\overline{)72}$ $\overset{8}{}$ ✓ $9\overline{)18}$ $\overset{3}{}$ $9\overline{)45}$ $\overset{5}{}$ ✓ $9\overline{)9}$ $\overset{81}{}$ ✗

$9\overline{)63}$ $\overset{7}{}$ ✓ $9\overline{)27}$ $\overset{3}{}$ $9\overline{)81}$ $\overset{9}{}$ $9\overline{)36}$ $\overset{4}{}$ ✓ $9\overline{)72}$ $\overset{8}{}$

34

Name _____

Divide. Check by multiplying.

$7\overline{)49}$ $\checkmark$ $8\overline{)16}$ $\checkmark$ $9\overline{)63}$ $\times$ $6\overline{)24}$ $\checkmark$ $7\overline{)21}$ $\checkmark$

$6\overline{)36}$ $\checkmark$ $8\overline{)64}$ $7\overline{)7}$ $\checkmark$ $9\overline{)27}$ $\checkmark$ $6\overline{)18}$ $\checkmark$

$7\overline{)56}$ $\checkmark$ $9\overline{)36}$ $\times$ $8\overline{)40}$ $\checkmark$ $7\overline{)28}$ $\checkmark$ $8\overline{)72}$ $\checkmark$

$9\overline{)45}$ $\checkmark$ $6\overline{)48}$ $\times$ $8\overline{)24}$ $\checkmark$ $7\overline{)42}$ $6\overline{)30}$ $\checkmark$

$9\overline{)9}$ $\times$ $8\overline{)32}$ $\checkmark$ $9\overline{)81}$ $\checkmark$ $6\overline{)54}$ $\checkmark$ $7\overline{)35}$ $\checkmark$

$6\overline{)42}$ $9\overline{)54}$ $\checkmark$ $7\overline{)63}$ $\times$ $8\overline{)56}$ $\checkmark$ $9\overline{)72}$ $\checkmark$

Name _____

Solve. Check each answer.

1. The Garcias took 63 jars to the glass collecting center. Each bag held 7 jars. How many bags of jars did they take in?

2. Carla works in a flower shop. She used 56 tulips in 8 different arrangements. How many tulips did she use in each arrangement?

3. Last weekend 64 people came to tour an old house. The guide can take only 8 people at a time. How many tours did she give?

4. A restaurant bill for $42 was shared equally by 7 people. How much did each person pay?

5. Linda earned $81 cutting the Newton's lawn this summer. If she cut the lawn 9 times, how much did she earn each time?

6. Hank used 54 screws to build a bookcase. He used 6 screws for each shelf. How many shelves does the bookcase have?

7. A store paid $72 to run a newspaper ad 9 times. How much did it cost to run the ad each time?

8. Barb used 49 sheets of paper to copy a report. If the report was 7 pages long, how many copies of the report did she make?

Problem Solving
© The Continental Press, Inc.

Name _____

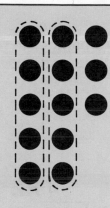

There are 2 groups of 5 in 13.

There are 3 left over.

$$5\overline{)13}$$ with 2R3 above
$$-10$$
$$3$$

$$13 \div 5 = 2 \text{ R}3$$

How many groups of 4 in 22? __5__

Circle them.

How many left over? __2__

$$22 \div 4 = \underline{5R2}$$

$$4\overline{)22}$$ with 5R2 above

How many groups of 3 in 14? __4__

Circle them.

How many left over? __2__

$$14 \div 3 = \underline{4R2}$$

$$3\overline{)14}$$ with 4R2 above

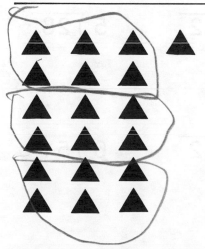

How many groups of 6 in 19? __3__

Circle them.

How many left over? __1__

$$19 \div 6 = \underline{3R1}$$

$$6\overline{)19}$$ with 3R1 above

Introduction to Division with Remainders
© The Continental Press, Inc.

Divide. (How many 3s in 14?)
$14 \div 3 = 4$
Write 4 in the quotient in the ones place.

Multiply.
$4 \times 3 = 12$
Write 12 under 14.

Subtract.
$14 - 12 = 2$
If the difference is less than the number you are dividing by, it is a remainder.
$2 < 3$

Write the remainder in the quotient.

$$3\overline{)14} \quad \overset{4}{\phantom{3\overline{)14}}} = \quad 3\overline{)14} \atop 12 = \quad 3\overline{)14} \atop \underline{-12} \atop 2 = \quad 3\overline{)14} \atop \underline{-12} \atop 2 \text{ R2}$$

Complete.

$$2\overline{)13} \quad 5\overline{)24} \quad 4\overline{)19} \quad 3\overline{)20} \quad 6\overline{)34}$$

Divide. Remember to write the remainder in the quotient.

$$7\overline{)46} \quad 2\overline{)19} \quad 5\overline{)39} \quad 6\overline{)15} \quad 9\overline{)74}$$

$$8\overline{)66} \quad 4\overline{)26} \quad 7\overline{)59} \quad 3\overline{)22} \quad 5\overline{)28}$$

$$9\overline{)79} \quad 5\overline{)47} \quad 4\overline{)30} \quad 7\overline{)62} \quad 6\overline{)51}$$

Basic Facts with Remainders
© The Continental Press, Inc.

$$\begin{array}{r} 7\ \text{R2} \\ 8\overline{)\ 58} \\ -\underline{56} \\ 2 \end{array}$$

Divide. $58 \div 8 = 7$
Multiply. $7 \times 8 = 56$
Subtract. $58 - 56 = 2$
Write the remainder R2 in the quotient.

Divide. Remember to write the remainder in the quotient.

$6\overline{)\ 49}$ $3\overline{)\ 8}$ $9\overline{)\ 59}$ $5\overline{)\ 23}$ $7\overline{)\ 67}$

$5\overline{)\ 38}$ $4\overline{)\ 13}$ $7\overline{)\ 36}$ $6\overline{)\ 16}$ $8\overline{)\ 58}$

$3\overline{)\ 14}$ $7\overline{)\ 25}$ $9\overline{)\ 88}$ $5\overline{)\ 44}$ $4\overline{)\ 27}$

$6\overline{)\ 46}$ $4\overline{)\ 35}$ $2\overline{)\ 15}$ $9\overline{)\ 47}$ $8\overline{)\ 69}$

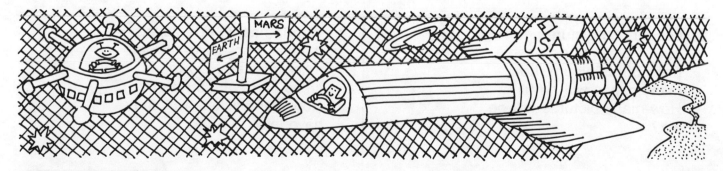

Name _____

Solve.

1. If 50 pancakes were shared equally by 7 people, how many pancakes did each person get? How many pancakes were left over?

2. A group of 32 people went to a nearby park. If 6 of them rode in each car, how many cars were filled? How many people rode in the extra car?

3. Jay has a collection of 71 rocks. If he puts 8 rocks in each display box, how many boxes will he fill? How many rocks will be left?

4. A jug holds 15 glasses of juice. If each person drinks 2 glasses of juice, how many people can be served? How many glasses will be left?

5. Nina's dogs eat 3 pounds of food a day. How long does a 25-pound bag of dog food last? How many pounds are left?

6. Mr. Ortega has 29 plants. If he plants 5 in each windowbox, how many boxes will he fill? How many plants will be left?

7. Les has 33 pictures. If he puts 4 pictures on each page of his album, how many pages will he fill? How many pictures will be left?

8. Wanda has 87 ride tickets. It costs 9 tickets to ride the roller coaster. How many times can Wanda ride it? How many tickets will she have left?

Problem Solving
© The Continental Press, Inc.

Name _____

Multiplying 10 is as easy as multiplying 1.

4 ones = 4

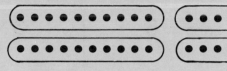

4 tens = 40

$$\begin{array}{r} 10 \\ \times 4 \\ \hline 40 \end{array}$$

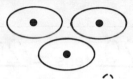

3 ones = 3

3 tens = 30

$$\begin{array}{r} 10 \\ \times 3 \\ \hline 30 \end{array}$$

Multiply.

7 tens = 70 $\begin{array}{r} 10 \\ \times 7 \\ \hline 70 \end{array}$    2 tens = ___ $\begin{array}{r} 10 \\ \times 2 \\ \hline \end{array}$

9 tens = ___ $\begin{array}{r} 10 \\ \times 9 \\ \hline \end{array}$ 5 tens = ___ $\begin{array}{r} 10 \\ \times 5 \\ \hline \end{array}$

6 tens = ___ $\begin{array}{r} 10 \\ \times 6 \\ \hline \end{array}$ 8 tens = ___ $\begin{array}{r} 10 \\ \times 8 \\ \hline \end{array}$

$\begin{array}{r} 10 \\ \times 6 \\ \hline \end{array}$ $\begin{array}{r} 10 \\ \times 4 \\ \hline \end{array}$ $\begin{array}{r} 10 \\ \times 9 \\ \hline \end{array}$ $\begin{array}{r} 10 \\ \times 3 \\ \hline \end{array}$ $\begin{array}{r} 10 \\ \times 1 \\ \hline \end{array}$ $\begin{array}{r} 10 \\ \times 2 \\ \hline \end{array}$ $\begin{array}{r} 10 \\ \times 8 \\ \hline \end{array}$ $\begin{array}{r} 10 \\ \times 7 \\ \hline \end{array}$

$\begin{array}{r} 10 \\ \times 5 \\ \hline \end{array}$ $\begin{array}{r} 5 \\ \times 10 \\ \hline \end{array}$ $\begin{array}{r} 10 \\ \times 7 \\ \hline \end{array}$ $\begin{array}{r} 7 \\ \times 10 \\ \hline \end{array}$ $\begin{array}{r} 10 \\ \times 9 \\ \hline \end{array}$ $\begin{array}{r} 9 \\ \times 10 \\ \hline \end{array}$ $\begin{array}{r} 10 \\ \times 4 \\ \hline \end{array}$ $\begin{array}{r} 4 \\ \times 10 \\ \hline \end{array}$

Multiplying 10
© The Continental Press, Inc.

41

Name _____

3×2 ones $= 6$ ones $= 6$

$$\begin{array}{r} 2 \\ \times 3 \\ \hline 6 \end{array}$$

3×2 tens $= 6$ tens $= 60$

$$\begin{array}{r} 20 \\ \times 3 \\ \hline 60 \end{array}$$

4×5 ones $= \underline{20}$ ones $= \underline{20}$

$$\begin{array}{r} 5 \\ \times 4 \\ \hline 20 \end{array}$$

4×5 tens $= \underline{20}$ tens $= \underline{200}$

$$\begin{array}{r} 50 \\ \times 4 \\ \hline 200 \end{array}$$

Multiply.

5×7 tens $= \underline{\hspace{1cm}}$ tens $= \underline{\hspace{1cm}}$

$$\begin{array}{r} 70 \\ \times 5 \\ \hline \end{array}$$

9×3 tens $= \underline{\hspace{1cm}}$ tens $= \underline{\hspace{1cm}}$

$$\begin{array}{r} 30 \\ \times 9 \\ \hline \end{array}$$

8×4 tens $= \underline{\hspace{1cm}}$ tens $= \underline{\hspace{1cm}}$

$$\begin{array}{r} 40 \\ \times 8 \\ \hline \end{array}$$

2×9 tens $= \underline{\hspace{1cm}}$ tens $= \underline{\hspace{1cm}}$

$$\begin{array}{r} 90 \\ \times 2 \\ \hline \end{array}$$

$$\begin{array}{r} 20 \\ \times 8 \\ \hline 160 \end{array} \qquad \begin{array}{r} 40 \\ \times 6 \\ \hline 0 \end{array} \qquad \begin{array}{r} 50 \\ \times 5 \\ \hline 0 \end{array} \qquad \begin{array}{r} 90 \\ \times 6 \\ \hline \end{array} \qquad \begin{array}{r} 30 \\ \times 2 \\ \hline \end{array} \qquad \begin{array}{r} 80 \\ \times 9 \\ \hline \end{array} \qquad \begin{array}{r} 70 \\ \times 3 \\ \hline \end{array}$$

$$\begin{array}{r} 30 \\ \times 4 \\ \hline \end{array} \qquad \begin{array}{r} 70 \\ \times 7 \\ \hline \end{array} \qquad \begin{array}{r} 40 \\ \times 9 \\ \hline \end{array} \qquad \begin{array}{r} 80 \\ \times 2 \\ \hline \end{array} \qquad \begin{array}{r} 50 \\ \times 8 \\ \hline \end{array} \qquad \begin{array}{r} 60 \\ \times 3 \\ \hline \end{array} \qquad \begin{array}{r} 20 \\ \times 5 \\ \hline \end{array}$$

Name _____

Multiplying 100 is as easy as multiplying 1 and 10.			
6 ones = 6	$\begin{array}{r} 1 \\ \times 6 \\ \hline 6 \end{array}$	6 tens = 60	$\begin{array}{r} 10 \\ \times 6 \\ \hline 60 \end{array}$ 6 hundreds = 600	$\begin{array}{r} 100 \\ \times 6 \\ \hline 600 \end{array}$

3 ones = _3_ $\begin{array}{r} 1 \\ \times 3 \\ \hline 3 \end{array}$ 3 tens = _30_ $\begin{array}{r} 10 \\ \times 3 \\ \hline 30 \end{array}$ 3 hundreds = _300_ $\begin{array}{r} 100 \\ \times 3 \\ \hline 300 \end{array}$

Multiply.

8 hundreds = _800_ $\begin{array}{r} 100 \\ \times 8 \\ \hline 800 \end{array}$ 5 hundreds = _____ $\begin{array}{r} 100 \\ \times 5 \\ \hline \end{array}$

4 hundreds = _____ $\begin{array}{r} 100 \\ \times 4 \\ \hline \end{array}$ 9 hundreds = _____ $\begin{array}{r} 100 \\ \times 9 \\ \hline \end{array}$

$\begin{array}{r} 100 \\ \times 7 \\ \hline \end{array}$ $\begin{array}{r} 100 \\ \times 2 \\ \hline \end{array}$ $\begin{array}{r} 100 \\ \times 6 \\ \hline \end{array}$ $\begin{array}{r} 100 \\ \times 4 \\ \hline \end{array}$ $\begin{array}{r} 100 \\ \times 3 \\ \hline \end{array}$

$\begin{array}{r} 10 \\ \times 4 \\ \hline \end{array}$ $\begin{array}{r} 1 \\ \times 9 \\ \hline \end{array}$ $\begin{array}{r} 100 \\ \times 5 \\ \hline \end{array}$ $\begin{array}{r} 10 \\ \times 7 \\ \hline \end{array}$ $\begin{array}{r} 100 \\ \times 8 \\ \hline \end{array}$

$\begin{array}{r} 100 \\ \times 2 \\ \hline \end{array}$ $\begin{array}{r} 10 \\ \times 8 \\ \hline \end{array}$ $\begin{array}{r} 1 \\ \times 8 \\ \hline \end{array}$ $\begin{array}{r} 10 \\ \times 3 \\ \hline \end{array}$ $\begin{array}{r} 100 \\ \times 9 \\ \hline \end{array}$

Name _____

Multiplying hundreds is as easy as multiplying ones and tens.

4×7 ones =	4×7 tens =	4×7 hundreds =
28 ones = 28	28 tens = 280	28 hundreds = 2800
$\begin{array}{r} 7 \\ \times 4 \\ \hline 28 \end{array}$	$\begin{array}{r} 70 \\ \times 4 \\ \hline 280 \end{array}$	$\begin{array}{r} 700 \\ \times 4 \\ \hline 2800 \end{array}$

Multiply.

6×3 hundreds =

_____ hundreds = _____

$$\begin{array}{r} 300 \\ \times 6 \\ \hline \end{array}$$

5×2 hundreds =

_____ hundreds = _____

$$\begin{array}{r} 200 \\ \times 5 \\ \hline \end{array}$$

$$\begin{array}{r} 400 \\ \times 2 \\ \hline 800 \end{array} \qquad \begin{array}{r} 300 \\ \times 7 \\ \hline 00 \end{array} \qquad \begin{array}{r} 800 \\ \times 4 \\ \hline 0 \end{array} \qquad \begin{array}{r} 500 \\ \times 8 \\ \hline \end{array} \qquad \begin{array}{r} 900 \\ \times 6 \\ \hline \end{array}$$

$$\begin{array}{r} 70 \\ \times 5 \\ \hline \end{array} \qquad \begin{array}{r} 900 \\ \times 2 \\ \hline \end{array} \qquad \begin{array}{r} 700 \\ \times 6 \\ \hline \end{array} \qquad \begin{array}{r} 40 \\ \times 8 \\ \hline \end{array} \qquad \begin{array}{r} 500 \\ \times 3 \\ \hline \end{array}$$

$$\begin{array}{r} 50 \\ \times 9 \\ \hline \end{array} \qquad \begin{array}{r} 200 \\ \times 6 \\ \hline \end{array} \qquad \begin{array}{r} 30 \\ \times 3 \\ \hline \end{array} \qquad \begin{array}{r} 400 \\ \times 2 \\ \hline \end{array} \qquad \begin{array}{r} 90 \\ \times 9 \\ \hline \end{array}$$

44

Multiplying Hundreds
© The Continental Press, Inc.

Name _____

Solve.

1. Carol got 3 packs of bolts at the hardware store. There are 10 bolts in each pack. How many bolts did she buy in all?

2. Mr. Heinz ordered 2 bags of cement mix. A bag weighs 80 pounds. How many pounds of cement mix did he buy in all?

3. Mr. Lum bought 5 giant packs of sandpaper. There were 20 sheets in each pack. How many sheets of sandpaper did Mr. Lum buy?

4. Ms. Montez bought 7 boxes of floor tiles. Each box contained 10 tiles. How many tiles did she buy altogether?

5. Mrs. Pitman bought 3 rolls of tape. There were 100 feet on each roll. How many feet of tape did Mrs. Pitman buy?

6. Mike bought 4 boxes of nails. There are 100 nails in a box. How many nails did he buy altogether?

7. Greg ordered 9 cases of shingles for his roof. A case holds 200 shingles. How many shingles did he order altogether?

8. Mr. Hess needed 6 rolls of electrical wiring. Each roll contains 500 feet. How many feet of wiring did he buy in all?

$$\begin{array}{c}34\\ \times 2\end{array} = \begin{array}{c}30\\ \times 2\\ \hline 60\end{array} + \begin{array}{c}4\\ \times 2\\ \hline 8\end{array} = \begin{array}{c}34\\ \times 2\\ \hline 68\end{array}$$

Complete.

$$\begin{array}{c}20\\ \times 6\end{array} + \begin{array}{c}1\\ \times 6\end{array} = \begin{array}{c}21\\ \times 6\end{array}$$
$$+ \qquad =$$

$$\begin{array}{c}80\\ \times 2\end{array} + \begin{array}{c}3\\ \times 2\end{array} = \begin{array}{c}83\\ \times 2\end{array}$$
$$+ \qquad =$$

$$\begin{array}{c}70\\ \times 2\end{array} + \begin{array}{c}2\\ \times 2\end{array} = \begin{array}{c}72\\ \times 2\end{array}$$
$$+ \qquad =$$

$$\begin{array}{c}40\\ \times 5\end{array} + \begin{array}{c}1\\ \times 5\end{array} = \begin{array}{c}41\\ \times 5\end{array}$$
$$+ \qquad =$$

$$\begin{array}{c}30\\ \times 3\end{array} + \begin{array}{c}3\\ \times 3\end{array} = \begin{array}{c}33\\ \times 3\end{array}$$
$$+ \qquad =$$

$$\begin{array}{c}60\\ \times 2\end{array} + \begin{array}{c}4\\ \times 2\end{array} = \begin{array}{c}64\\ \times 2\end{array}$$
$$+ \qquad =$$

$$\begin{array}{c}40\\ \times 4\end{array} + \begin{array}{c}2\\ \times 4\end{array} = \begin{array}{c}42\\ \times 4\end{array}$$
$$+ \qquad =$$

$$\begin{array}{c}50\\ \times 3\end{array} + \begin{array}{c}2\\ \times 3\end{array} = \begin{array}{c}52\\ \times 3\end{array}$$
$$+ \qquad =$$

$$\begin{array}{c}90\\ \times 7\end{array} + \begin{array}{c}1\\ \times 7\end{array} = \begin{array}{c}91\\ \times 7\end{array}$$
$$+ \qquad =$$

$$\begin{array}{c}20\\ \times 2\end{array} + \begin{array}{c}4\\ \times 2\end{array} = \begin{array}{c}24\\ \times 2\end{array}$$
$$+ \qquad =$$

YO-YO COMPETITION

Multiplying 1 Digit × 2 Digits
© The Continental Press, Inc.

	Multiply the ones.	Multiply the tens.	Add.

$$\begin{array}{r} 32 \\ \times 3 \\ \hline \end{array} =$$

$$\begin{array}{r} 32 \\ \times 3 \\ \hline 6 \end{array} \ (3 \times 2)$$

$$\begin{array}{r} 32 \\ \times 3 \\ \hline 6 \\ 90 \end{array} \ (3 \times 30)$$

$$\begin{array}{r} 32 \\ \times 3 \\ \hline 6 \\ 90 \\ \hline 96 \end{array}$$

Multiply.

$$\begin{array}{r} 54 \\ \times 2 \\ \hline 8 \\ 100 \\ \hline 108 \end{array} \begin{array}{l} (2 \times 4) \\ (2 \times 50) \end{array}$$

$$\begin{array}{r} 81 \\ \times 6 \\ \hline \end{array} \begin{array}{l} (6 \times 1) \\ (6 \times 80) \end{array}$$

$$\begin{array}{r} 23 \\ \times 2 \\ \hline \end{array} \begin{array}{l} (2 \times 3) \\ (2 \times 20) \end{array}$$

$$\begin{array}{r} 12 \\ \times 4 \\ \hline 8 \\ 40 \\ \hline 48 \end{array} \qquad \begin{array}{r} 42 \\ \times 3 \\ \hline \end{array} \qquad \begin{array}{r} 28 \\ \times 1 \\ \hline \end{array} \qquad \begin{array}{r} 74 \\ \times 2 \\ \hline \end{array} \qquad \begin{array}{r} 63 \\ \times 3 \\ \hline \end{array} \qquad \begin{array}{r} 51 \\ \times 6 \\ \hline \end{array} \qquad \begin{array}{r} 24 \\ \times 2 \\ \hline \end{array}$$

$$\begin{array}{r} 71 \\ \times 5 \\ \hline \end{array} \qquad \begin{array}{r} 31 \\ \times 8 \\ \hline \end{array} \qquad \begin{array}{r} 54 \\ \times 2 \\ \hline \end{array} \qquad \begin{array}{r} 13 \\ \times 3 \\ \hline \end{array} \qquad \begin{array}{r} 82 \\ \times 4 \\ \hline \end{array} \qquad \begin{array}{r} 62 \\ \times 2 \\ \hline \end{array} \qquad \begin{array}{r} 81 \\ \times 7 \\ \hline \end{array}$$

Name _____

	Multiply the ones.	Multiply the tens.	Multiply the hundreds.	Add.

$$432 \atop \underline{\times 3}$$ =

$$\begin{array}{r} 43\mathbf{2} \\ \underline{\times 3} \\ \mathbf{6} \ (2 \times 3) \end{array}$$

$$\begin{array}{r} 432 \\ \underline{\times 3} \\ 6 \\ \mathbf{90} \ (3 \times 30) \end{array}$$

$$\begin{array}{r} 432 \\ \underline{\times 3} \\ 6 \\ 90 \\ \underline{\mathbf{1200}} \ (3 \times 400) \end{array}$$

$$\begin{array}{r} 432 \\ \underline{\times 3} \\ \mathbf{6} \\ \mathbf{90} \\ \underline{\mathbf{1200}} \\ \mathbf{1296} \end{array}$$

Multiply.

$$\begin{array}{r} 134 \\ \underline{\times 2} \\ 8 \ (2 \times 4) \\ 60 \ (2 \times 30) \\ \underline{200} \ (2 \times 100) \\ 268 \end{array}$$

$$\begin{array}{r} 312 \\ \underline{\times 3} \\ (3 \times 2) \\ (3 \times 10) \\ \underline{} \ (3 \times 300) \end{array}$$

$$\begin{array}{r} 510 \\ \underline{\times 4} \\ (4 \times 0) \\ (4 \times 10) \\ \underline{} \ (4 \times 500) \end{array}$$

$$\begin{array}{r} 543 \\ \underline{\times 2} \\ 6 \\ 80 \\ \underline{1000} \end{array}$$

$$221 \atop \underline{\times 3}$$

$$642 \atop \underline{\times 2}$$

$$911 \atop \underline{\times 8}$$

$$417 \atop \underline{\times 1}$$

$$723 \atop \underline{\times 3}$$

$$814 \atop \underline{\times 2}$$

$$413 \atop \underline{\times 3}$$

$$931 \atop \underline{\times 3}$$

$$520 \atop \underline{\times 2}$$

$$112 \atop \underline{\times 4}$$

$$420 \atop \underline{\times 4}$$

48

Solve.

1. Mrs. Woburn put 3 gallons of gas in her car. If her car gets 23 miles per gallon, how far can she go?

2. The Bowman's motor home gets only 7 miles to a gallon of gas. How far can they travel on 60 gallons of gas?

3. Antonio's small car gets 31 miles per gallon of gas. How far can he drive on 9 gallons of gas?

4. Lucy put 3 gallons of gas in her motorcycle. If she gets 52 miles to a gallon, how far can she go?

5. Joan's car gets 20 miles per gallon on the highway. How many miles can she travel on 8 gallons?

6. Tom sold oil to 134 people last week. If each person bought 2 quarts of oil, how many quarts of oil did Tom sell altogether?

7. During the winter sale, Stan's Garage sold 62 packs of windshield wiper blades. There are 2 wiper blades in each pack. How many blades were sold in all?

8. A tractor trailer filled its tanks with 200 gallons of diesel fuel. It gets only 5 miles to a gallon. How far can it go on 200 gallons?

Name _____

	Multiply the ones.		Multiply the tens.		Add.
63 ×5	=	63 ×5 **15**	=	63 ×5 **15** **300**	=	63 ×5 **15** **300** **315**

Multiply.

76 ×3 18 210 228	24 ×7	39 ×2	46 ×6	82 ×5	95 ×8

58 ×4	99 ×5	42 ×9	84 ×3	33 ×7	96 ×6

27 ×8	45 ×4	37 ×7	82 ×6	97 ×9	65 ×3

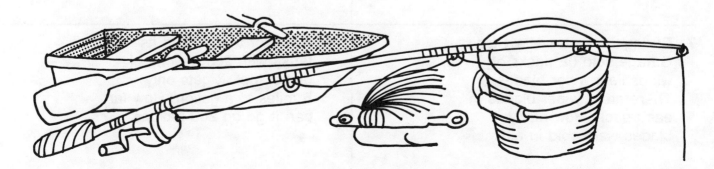

Multiplying 1 Digit × 2 Digits with Regrouping

Name _____

There is a shorter way to multiply.

Multiply the ones.
$6 \times 4 = 24$
Write the 4 ones.
Remember the
2 tens.

Multiply the tens.
$6 \times 5 = 30$
Add the 2 tens.
$30 + 2 = 32$

$$\begin{array}{r} 54 \\ \times 6 \end{array} \quad = \quad \begin{array}{r} 2 \\ 54 \\ \times 6 \\ \hline 4 \end{array} \quad = \quad \begin{array}{r} 2 \\ 54 \\ \times 6 \\ \hline 324 \end{array} \quad = \quad \begin{array}{r} 2 \\ 54 \\ \times 6 \\ \hline 324 \end{array}$$

Complete.

$$\begin{array}{r} 5 \\ 48 \\ \times 7 \\ \hline 336 \end{array} \qquad \begin{array}{r} 4 \\ 25 \\ \times 8 \\ \hline 0 \end{array} \qquad \begin{array}{r} 1 \\ 33 \\ \times 5 \\ \hline 5 \end{array} \qquad \begin{array}{r} 3 \\ 79 \\ \times 4 \\ \hline 6 \end{array} \qquad \begin{array}{r} 1 \\ 92 \\ \times 9 \\ \hline 8 \end{array} \qquad \begin{array}{r} 4 \\ 67 \\ \times 6 \\ \hline 2 \end{array}$$

Multiply. Use the shorter way.

$$\begin{array}{r} 2 \\ 16 \\ \times 4 \\ \hline 64 \end{array} \qquad \begin{array}{r} 82 \\ \times 9 \end{array} \qquad \begin{array}{r} 63 \\ \times 8 \end{array} \qquad \begin{array}{r} 46 \\ \times 6 \end{array} \qquad \begin{array}{r} 29 \\ \times 3 \end{array} \qquad \begin{array}{r} 95 \\ \times 5 \end{array}$$

$$\begin{array}{r} 37 \\ \times 3 \end{array} \qquad \begin{array}{r} 42 \\ \times 8 \end{array} \qquad \begin{array}{r} 58 \\ \times 6 \end{array} \qquad \begin{array}{r} 49 \\ \times 4 \end{array} \qquad \begin{array}{r} 34 \\ \times 9 \end{array} \qquad \begin{array}{r} 87 \\ \times 7 \end{array}$$

$$\begin{array}{r} 93 \\ \times 7 \end{array} \qquad \begin{array}{r} 38 \\ \times 8 \end{array} \qquad \begin{array}{r} 96 \\ \times 9 \end{array} \qquad \begin{array}{r} 83 \\ \times 6 \end{array} \qquad \begin{array}{r} 55 \\ \times 4 \end{array} \qquad \begin{array}{r} 28 \\ \times 5 \end{array}$$

$$\begin{array}{r} 54 \\ \times 2 \\ \hline 108 \end{array} \qquad \begin{array}{r} {}^{4} \\ 76 \\ \times 7 \\ \hline 532 \end{array}$$

Multiply. Use the shorter way.

$$\begin{array}{r} {}^{3} \\ 35 \\ \times 6 \\ \hline 210 \end{array} \qquad \begin{array}{r} 19 \\ \times 7 \\ \hline \end{array} \qquad \begin{array}{r} 68 \\ \times 8 \\ \hline \end{array} \qquad \begin{array}{r} 83 \\ \times 2 \\ \hline \end{array} \qquad \begin{array}{r} 25 \\ \times 5 \\ \hline \end{array} \qquad \begin{array}{r} 40 \\ \times 3 \\ \hline \end{array}$$

$$\begin{array}{r} 63 \\ \times 3 \\ \hline \end{array} \qquad \begin{array}{r} 29 \\ \times 6 \\ \hline \end{array} \qquad \begin{array}{r} 37 \\ \times 1 \\ \hline \end{array} \qquad \begin{array}{r} 48 \\ \times 9 \\ \hline \end{array} \qquad \begin{array}{r} 65 \\ \times 7 \\ \hline \end{array} \qquad \begin{array}{r} 54 \\ \times 4 \\ \hline \end{array}$$

$$\begin{array}{r} 38 \\ \times 4 \\ \hline \end{array} \qquad \begin{array}{r} 73 \\ \times 9 \\ \hline \end{array} \qquad \begin{array}{r} 54 \\ \times 3 \\ \hline \end{array} \qquad \begin{array}{r} 70 \\ \times 8 \\ \hline \end{array} \qquad \begin{array}{r} 39 \\ \times 2 \\ \hline \end{array} \qquad \begin{array}{r} 87 \\ \times 5 \\ \hline \end{array}$$

$$\begin{array}{r} 71 \\ \times 5 \\ \hline \end{array} \qquad \begin{array}{r} 81 \\ \times 6 \\ \hline \end{array} \qquad \begin{array}{r} 43 \\ \times 8 \\ \hline \end{array} \qquad \begin{array}{r} 27 \\ \times 9 \\ \hline \end{array} \qquad \begin{array}{r} 74 \\ \times 7 \\ \hline \end{array} \qquad \begin{array}{r} 62 \\ \times 4 \\ \hline \end{array}$$

Multiplying 1 Digit × 2 Digits with and without Regrouping

Name _____

Name

Solve.

1. Mindy used 5 rolls of film on her trip to Hawaii. If each roll has 20 pictures on it, how many pictures did she take altogether?

2. Don picked out 11 of his favorite pictures. He had 4 extra copies made of each picture. How many extra copies did he have made in all?

3. Beth bought 2 rolls of film at the photo shop. There are 24 pictures on each roll. How many pictures can Beth take in all?

4. Parkside Photo Shop sells about 3 cameras a week. How many cameras would the shop sell in 52 weeks?

5. John bought 7 packs of flashcubes last month. Each pack had 12 cubes in it. How many flashcubes did John buy?

6. A newspaper photographer shot 9 rolls of film last week. Each roll had 36 pictures on it. How many pictures did he shoot in all?

7. Nicola worked at Parkside Photo Shop last summer. She worked 8 hours a day for a total of 74 days. How many hours did she work in all?

8. Parkside Photo Shop sells about 45 rolls of film a day. How many rolls would the shop sell in 6 days?

Name _____

	Multiply the ones.	Multiply the tens.	Multiply the hundreds.	Add.
325 ×7 =	325 ×7 **35** =	325 ×7 35 **140** =	**325** **×7** 35 140 **2100** =	**325** **×7** **35** **140** **2100** **2275**

Multiply.

$$\begin{array}{r} 932 \\ \times 6 \\ \hline 12 \\ 180 \\ 5400 \\ \hline 5592 \end{array}$$

$$\begin{array}{r} 415 \\ \times 8 \\ \hline \end{array}$$

$$\begin{array}{r} 273 \\ \times 9 \\ \hline \end{array}$$

$$\begin{array}{r} 194 \\ \times 4 \\ \hline \end{array}$$

$$\begin{array}{r} 746 \\ \times 3 \\ \hline \end{array}$$

$$\begin{array}{r} 297 \\ \times 5 \\ \hline \end{array}$$

$$\begin{array}{r} 684 \\ \times 6 \\ \hline \end{array}$$

$$\begin{array}{r} 855 \\ \times 2 \\ \hline \end{array}$$

$$\begin{array}{r} 379 \\ \times 3 \\ \hline \end{array}$$

$$\begin{array}{r} 368 \\ \times 2 \\ \hline \end{array}$$

$$\begin{array}{r} 943 \\ \times 9 \\ \hline \end{array}$$

$$\begin{array}{r} 482 \\ \times 7 \\ \hline \end{array}$$

$$\begin{array}{r} 256 \\ \times 8 \\ \hline \end{array}$$

$$\begin{array}{r} 535 \\ \times 5 \\ \hline \end{array}$$

Multiplying 1 Digit × 3 Digits with Regrouping

Name _____

Complete.

$$
\begin{array}{r} {}^{2}329 \\ \times 3 \\ \hline 987 \end{array}
\qquad
\begin{array}{r} {}^{6}508 \\ \times 8 \\ \hline 4 \end{array}
\qquad
\begin{array}{r} {}^{3}415 \\ \times 6 \\ \hline 0 \end{array}
\qquad
\begin{array}{r} {}^{1}627 \\ \times 2 \\ \hline 4 \end{array}
\qquad
\begin{array}{r} {}^{2}514 \\ \times 5 \\ \hline 0 \end{array}
$$

Multiply. Use the shorter way.

$$
\begin{array}{r} {}^{3}904 \\ \times 9 \\ \hline 8136 \end{array}
\qquad
\begin{array}{r} 547 \\ \times 2 \\ \hline \end{array}
\qquad
\begin{array}{r} 506 \\ \times 6 \\ \hline \end{array}
\qquad
\begin{array}{r} 905 \\ \times 8 \\ \hline \end{array}
\qquad
\begin{array}{r} 825 \\ \times 3 \\ \hline \end{array}
$$

$$
\begin{array}{r} 638 \\ \times 2 \\ \hline \end{array}
\qquad
\begin{array}{r} 717 \\ \times 5 \\ \hline \end{array}
\qquad
\begin{array}{r} 424 \\ \times 4 \\ \hline \end{array}
\qquad
\begin{array}{r} 813 \\ \times 6 \\ \hline \end{array}
\qquad
\begin{array}{r} 309 \\ \times 9 \\ \hline \end{array}
$$

$$
\begin{array}{r} 823 \\ \times 4 \\ \hline \end{array}
\qquad
\begin{array}{r} 729 \\ \times 3 \\ \hline \end{array}
\qquad
\begin{array}{r} 314 \\ \times 6 \\ \hline \end{array}
\qquad
\begin{array}{r} 704 \\ \times 7 \\ \hline \end{array}
\qquad
\begin{array}{r} 107 \\ \times 5 \\ \hline \end{array}
$$

Name _____

Do you remember the shorter way to multiply?

Multiply the ones. $3 \times 2 = 6$		Multiply the tens. $3 \times 5 = 15$ Write the 5 tens. Remember the 1 hundred.		Multiply the hundreds. $3 \times 4 = 12$ Add the 1 hundred. $12 + 1 = 13$		
452 ×3 ── 6	=	1 452 ×3 ── 56	=	1 452 ×3 ──── 1356	=	1 452 ×3 ──── 1356

Complete.

² 872 ×4 ──── 3488	³ 361 ×6 ──── 66	⁶ 470 ×9 ──── 30	¹ 253 ×3 ──── 59	² 931 ×8 ──── 48

Multiply. Use the shorter way.

¹ 382 ×2 ──── 764	790 ×6	521 ×5	820 ×8	643 ×3

930 ×9	651 ×8	441 ×7	340 ×6	762 ×4

580 ×7	262 ×3	763 ×2	972 ×4	693 ×3

56

Multiplying 1 Digit × 3 Digits, Regrouping Tens
© The Continental Press, Inc.

Name _____

```
       3              2
      317            462
      ×5             ×4
    ─────          ─────
     1585           1848
```

Multiply. Use the shorter way.

```
  563          915          393          609
  ×3           ×6           ×2           ×7
 ─────        ─────        ─────        ─────
 1689
```

```
  712          541          205          418
  ×8           ×5           ×9           ×4
 ─────        ─────        ─────        ─────
```

```
  354          260          414          923
  ×2           ×6           ×7           ×4
 ─────        ─────        ─────        ─────
```

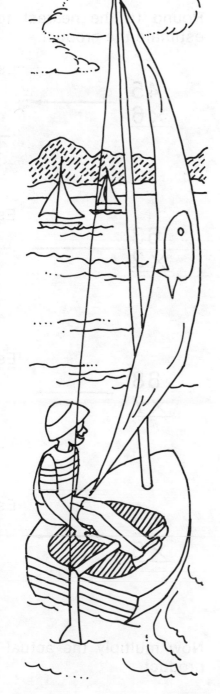

```
  809          983          791          627
  ×5           ×3           ×9           ×2
 ─────        ─────        ─────        ─────
```

```
  312          124          471          213
  ×8           ×4           ×6           ×7
 ─────        ─────        ─────        ─────
```

Name _____

An estimated answer is a guess near the actual answer. To estimate a product, round the number being multiplied to the nearest ten or hundred. Then multiply the rounded number.

	Estimate			Estimate
56	60		823	800
×7	×7		×4	×4
	420			3200

Round to the nearest ten and estimate the product.

Round to the nearest hundred and estimate the product.

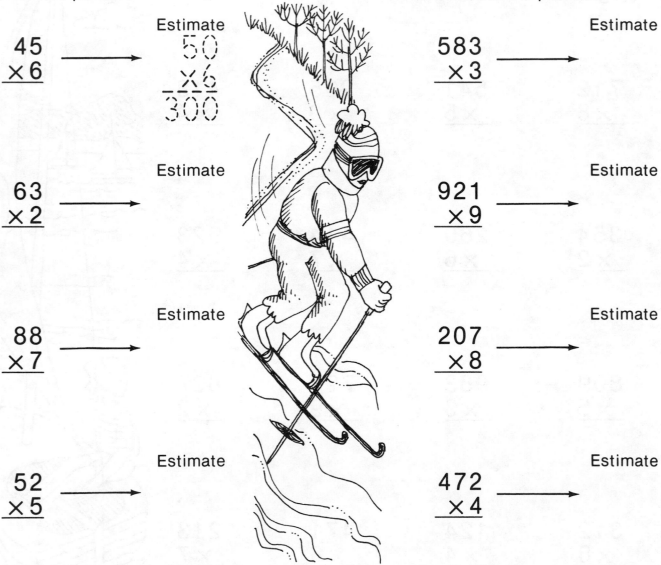

45
×6 ⟶ Estimate
          50
          ×6
          300

583
×3 ⟶ Estimate

63
×2 ⟶ Estimate

921
×9 ⟶ Estimate

88
×7 ⟶ Estimate

207
×8 ⟶ Estimate

52
×5 ⟶ Estimate

472
×4 ⟶ Estimate

Now multiply the actual numbers. Compare your estimated product with the actual product.

Estimating Products
© The Continental Press, Inc.

Name _____

1. On a TV program a man lifted
   2 weights. Each one weighed
   245 pounds. How much did he
   lift altogether?

2. The fish farm has 9 tanks with
   580 fish in each. How many fish
   does the farm have in all?

3. Ms. Moss guides 115 tourists
   each day. How many people
   does she guide in 6 days?

4. Marcus wrote a 3-page report. If
   there were about 225 words on
   each page, how many words
   were in the entire report?

5. Mrs. Watanabe made a 471-mile
   business trip 5 times this year.
   How far did she travel in all?

6. Carlos unloaded 8 new washing
   machines. Each one weighed
   228 pounds. How many pounds
   in all did he unload?

7. An adult elephant eats about
   150 pounds of food a day. How
   much food would it eat in 7 days?

8. A jet travels 631 miles per hour.
   How far will it travel in 6 hours?

Name _____

Sometimes you must regroup more than once.

Multiply the ones and regroup.		Multiply the tens and regroup.		Multiply the hundreds.		
3 **265** **×7** **5**	=	**4 3** **265** **×7** **55**	=	**4 3** **265** **×7** **1855**	=	4 3 265 ×7 1855

Multiply.

5 1
472
×8
3776

963
×4

639
×3

284
×5

766
×2

336
×5

673
×9

845
×6

329
×4

475
×7

587
×2

237
×7

458
×3

736
×8

639
×5

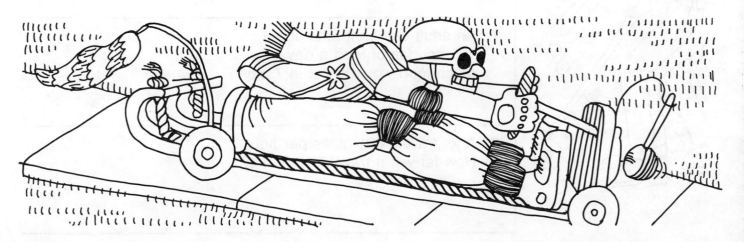

Multiplying 1 Digit × 3 Digits, Regrouping Twice
© The Continental Press, Inc.

Name _____

Multiply dollars and cents the same way you multiply other numbers. Regroup if necessary. Remember to write the dollar sign and decimal point in the answer.

	2	3 4
$3.14	$5.73	$6.58
×2	×3	×6
$6.28	$17.19	$39.48

Multiply.

$9.67
×4
$38.68

$5.82
×5

$3.98
×3

$8.69
×8

$1.39
×6

$7.56
×2

$4.27
×7

$2.25
×4

$6.98
×9

$2.07
×8

$4.31
×5

$9.88
×2

$7.20
×6

$3.60
×4

$8.75
×2

$5.42
×3

Name _____

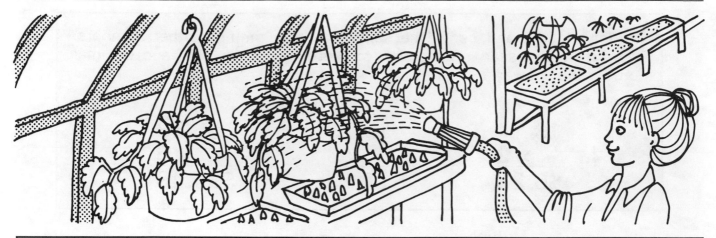

Solve.

1. Jody works at Berg's Greenhouse for 3 hours each evening. She is paid $3.45 an hour. How much does she earn in an evening?

2. A rancher bought 9 rolls of wire fencing. Each roll is 124 feet long. How many feet of fencing did the rancher buy in all?

3. Every weekday Chuck delivers 624 newspapers. How many newspapers does he deliver in 5 days?

4. Today Mrs. Suarez bought 7 window shades. Each one cost $8.95. How much did she pay for the shades in all?

5. Missy works in a hotel kitchen. Today she cut 487 grapefruit into 2 halves each. How many grapefruit halves did she have?

6. On Saturday Mr. Kaufman bought 4 ferns at $9.35 each. How much did he pay altogether?

7. The Ozawa family bought 8 bags of wild bird seed this winter. Each bag cost $5.25. How much did the seed cost altogether?

8. A train ticket to Boston costs $7.75 per person. How much would tickets cost a family of 6 people?

Multiply the ones.		Multiply the tens.		Multiply the hundreds. $2 \times 6 = 12$ Write the 2 hundreds. Remember the 1 thousand.		Multiply the thousands. $2 \times 4 = 8$ Add the 1 thousand.		
4623 $\times 2$ ——— 6	=	4623 $\times 2$ ——— 46	=	$\overset{1}{4}$623 $\times 2$ ——— 246	=	$\overset{1}{4}$623 $\times 2$ ——— 9246	=	4623 $\times 2$ ——— 9246

Complete.

$\overset{2}{2}310$
$\times 7$
————
16,170

$\overset{1}{7}914$
$\times 2$
————
828

$\overset{3}{5}822$
$\times 4$
————
288

$\overset{4}{6}901$
$\times 5$
————
505

$\overset{2}{3}731$
$\times 3$
————
193

Multiply.

$\overset{3}{4}921$
$\times 4$
————
19,684

7710
$\times 6$
————

5823
$\times 2$
————

9611
$\times 8$
————

3523
$\times 3$
————

8834
$\times 2$
————

6410
$\times 8$
————

2911
$\times 9$
————

7612
$\times 4$
————

4901
$\times 5$
————

9611
$\times 7$
————

3810
$\times 5$
————

8632
$\times 3$
————

6542
$\times 2$
————

7611
$\times 6$
————

Multiplying 1 Digit × 4 Digits, Regrouping Hundreds
© The Continental Press, Inc.

Name _____

Multiply.

```
  2 2
  3715        5980        4498        9077        1392
  × 4         × 6         × 2         × 8         × 3
 14,860
```

```
  6381        2829        7631        8608        5125
  × 7         × 3         × 5         × 9         × 4
```

```
  4805        1245        9530        3854        6713
  × 6         × 4         × 8         × 2         × 7
```

```
  6189        7941        5846        2915        8780
  × 3         × 9         × 2         × 6         × 5
```

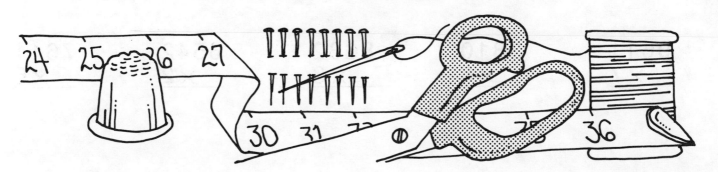

64

Multiplying 1 Digit × 4 Digits, Regrouping Twice
© The Continental Press, Inc.

Name _____

Multiply.

```
 1 3 3
 4276       6842       5537       1638       2953
  ×5         ×9         ×6         ×3         ×5
─────
21,380
```

```
 7685       3284       9845       4926       8415
  ×2         ×7         ×8         ×4         ×9
```

```
 5357       6734       1937       2597       7728
  ×4         ×3         ×5         ×2         ×6
```

```
 3276       8193       4269       5875       9517
  ×9         ×6         ×8         ×4         ×7
```

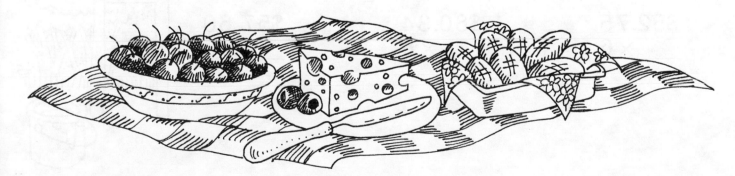

Multiplying 1 Digit × 4 Digits, Regrouping Three Times
© The Continental Press, Inc.

65

Name _____

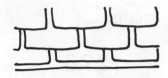

Multiply. Regroup as often as necessary.
Remember to write the dollar sign and
decimal point in the answer.

```
  1 1
$34.52
 × 3
$103.56
```

Multiply.

```
  3 1 3
$59.38        $61.89        $37.50
 × 4           × 8           × 3
$237.52
```

```
$86.17        $29.95        $94.82
 × 5           × 2           × 7
```

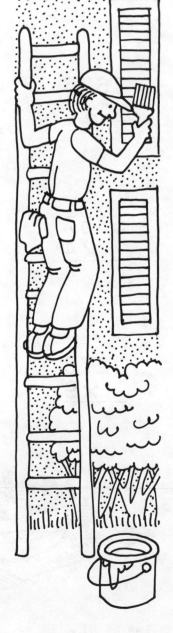

```
$64.50        $48.29        $75.13
 × 9           × 3           × 4
```

```
$52.75        $80.34        $57.60
 × 6           × 8           × 5
```

66

Multiplying Dollars and Cents

Name _____

Solve.

1. Harry earns $42.05 a day at Mid-Town Bakery. How much does he earn in 5 days?

2. For 2 days each week, Harry bakes rolls. Each day he bakes 6558 rolls. How many does he bake in 2 days?

3. One model of swimming pool was on sale for $5135 at Mill's Pool Shop. Last week Ms. Mill sold 7 of them. What were the total sales?

4. A bus ticket costs Fran $15.20 a month. How much does she spend on bus tickets in 9 months?

5. Crown Candy Company uses 2940 pounds of chocolate each week. How much does it use in 3 weeks?

6. The Message Makers printed 8255 T-shirts a day for 6 days. How many T-shirts were printed in all?

7. Once a week, Mr. Lightfoot prints 9325 copies of *Indian Ways*, a newspaper. How many copies does he print in 4 weeks?

8. Rob runs a machine that makes 3112 paper cups in an hour. How many cups does it make in 8 hours?

Name _____

Multiply to find the total cost of several items with the same price.
The @ sign means "each," as in "2 packs of cookies @ $.89."

$$\begin{array}{r} \$.89 \\ \times 2 \\ \hline \$1.78 \end{array} = \text{total cost}$$

Find the total cost of each group of things.

6 oranges @ $.15	2 bars of soap @ $.33	4 packs frozen onion rings @ $.61
$$\begin{array}{r} \$.15 \\ \times 6 \\ \hline \$.90 \end{array}$$		
2 jars of peanut butter @ $1.49	3 tablets @ $1.19	7 pounds of ground beef @ $1.65 a pound
9 cans of soup @ $.35	8 pounds of potatoes @ $.19 a pound	5 boxes of lightbulbs @ $1.50

Multiplication: Computing Prices

Name _____

Dividing tens is as easy as dividing ones. Remember to put a 0 in the ones place.

6 ones $\div 2 = 3$ ones	6 tens $\div 2 = 3$ tens
$6 \div 2 = 3$	$60 \div 2 = 30$
$2\overline{)6}^{\,3}$	$2\overline{)60}^{\,30}$

Divide.

$3\overline{)9}^{\,3}$ $\quad$ $3\overline{)90}^{\,30}$   $\quad$ $4\overline{)4}$ $\quad$ $4\overline{)40}$

$1\overline{)7}$ $\quad$ $1\overline{)70}$ $\quad\quad\quad\quad\quad$ $3\overline{)6}$ $\quad$ $3\overline{)60}$

$4\overline{)80}$ $\quad$ $2\overline{)20}$ $\quad$ $1\overline{)50}$ $\quad$ $9\overline{)90}$ $\quad$ $6\overline{)60}$

$2\overline{)40}$ $\quad$ $5\overline{)50}$ $\quad$ $8\overline{)80}$ $\quad$ $2\overline{)60}$ $\quad$ $1\overline{)80}$

$7\overline{)70}$ $\quad$ $1\overline{)90}$ $\quad$ $2\overline{)80}$ $\quad$ $1\overline{)40}$ $\quad$ $3\overline{)30}$

Name _____

Divide.

2)64   4)88   3)63   2)24

2)42   3)96   4)84   3)66

4)48   2)86   3)36   2)88

2)68   4)80   3)69   2)22

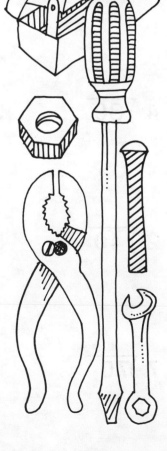

70

Dividing by 1 Digit for 2-Digit Quotients, No Regrouping
© The Continental Press, Inc.

Name _____

```
    41              41
2 ) 82            × 2
   -8              82
    02
   -2
     0
```

Divide. Check by multiplying.

```
    32
3 ) 96          ×  3
   -9             96
    6
   -6
```

```
2 ) 62          × _____
```

```
5 ) 50          × _____
```

```
4 ) 84          × _____
```

```
2 ) 68          × _____
```

```
3 ) 39          × _____
```

```
4 ) 48          × _____
```

```
7 ) 77          × _____
```

Name _____

Solve.

1. Ada sold 9 plant hangers at the crafts fair. She made $90. How much did she charge for each plant hanger?

2. Bruce has 84 pounds of clay. If he uses 4 pounds for each pot, how many pots can he make?

3. There are 44 light bulbs on a shelf at a store. If they come in packs of 4, how many packs are on the shelf?

4. Sally opened 69 clams. If she opened 3 clams a minute, how long did it take her to open all of them?

5. Julie has 82 balls of yarn. She uses 2 balls to weave a placemat. How many placemats can she make in all?

6. Carl can do 5 push-ups in a minute. How long would it take him to do 55 push-ups?

7. A restaurant manager bought 63 flowers. If he puts a vase with 3 flowers at each table, how many tables can he decorate?

8. Mr. Ace bought 48 eggs. If they came in 4 boxes, how many eggs were in each box?

Problem Solving
© The Continental Press, Inc.

Name _____

Divide the tens.
(How many 4s in 9?)
Multiply. $2 \times 4 = 8$
Subtract. $9 - 8 = 1$

$$\begin{array}{r} 2 \\ 4\overline{)96} \\ -8 \\ \hline 1 \end{array}$$

There is 1 ten left.
It cannot be divided by 4.
Think of it as 10 ones.
Bring down the 6 ones
and regroup.

$$\begin{array}{r} 2 \\ 4\overline{)96} \\ -8\downarrow \\ \hline 16 \end{array}$$

Divide the ones.
(How many 4s in 16?)
Multiply. $4 \times 4 = 16$
Subtract. $16 - 16 = 0$

$$\begin{array}{r} 24 \\ 4\overline{)96} \\ -8\downarrow \\ \hline 16 \\ -16 \\ \hline 0 \end{array}$$

Divide.

$3\overline{)78}$ $2\overline{)94}$ $4\overline{)56}$ $6\overline{)96}$ $5\overline{)75}$

$2\overline{)76}$ $4\overline{)60}$ $3\overline{)81}$ $5\overline{)65}$ $7\overline{)98}$

$6\overline{)84}$ $2\overline{)58}$ $7\overline{)84}$ $3\overline{)45}$ $2\overline{)52}$

$8\overline{)96}$ $5\overline{)80}$ $3\overline{)87}$ $6\overline{)90}$ $4\overline{)68}$

$$3\overline{)87}$$
29
−6↓
27
−27
0

Divide the tens.
Multiply. Then subtract.
Bring down the ones and regroup.
Divide the ones.
Multiply. Then subtract.

Divide.

$$4\overline{)56}$$ $$7\overline{)91}$$ $$5\overline{)85}$$ $$3\overline{)78}$$ $$6\overline{)84}$$

$$5\overline{)90}$$ $$2\overline{)98}$$ $$3\overline{)57}$$ $$4\overline{)72}$$ $$8\overline{)96}$$

$$7\overline{)84}$$ $$4\overline{)64}$$ $$2\overline{)70}$$ $$7\overline{)98}$$ $$6\overline{)72}$$

$$7\overline{)70}$$ $$4\overline{)92}$$ $$6\overline{)78}$$ $$3\overline{)84}$$ $$4\overline{)52}$$

Dividing by 1 Digit for 2-Digit Quotients, with Regrouping

Name _____

Divide. Remember to write the remainder in the quotient.

```
   23 R2
4 ) 94        5 ) 84        2 ) 79        8 ) 99        6 ) 87
   8
   14
   12
    2
```

```
5 ) 68        6 ) 73        3 ) 86        2 ) 95        7 ) 90
```

```
8 ) 97        4 ) 78        2 ) 53        7 ) 95        3 ) 76
```

Dividing by 1 Digit for 2-Digit Quotients, with Regrouping and Remainders
© The Continental Press, Inc.

Name _____

Find the quotient
by dividing.

$$3 \overline{)89} \quad \begin{array}{r} 29 \text{ R2} \\ \hline -6\downarrow \\ \hline 29 \\ -27 \\ \hline 2 \end{array}$$

$$\begin{array}{r} 29 \\ \times 3 \\ \hline 87 \\ + 2 \\ \hline 89 \end{array}$$

Check by multiplying,
then adding the remainder.
The answer should be the
same number that you
divided.

Divide and check.

$$5 \overline{)73} \quad \begin{array}{r} 14 \text{ R3} \\ \hline 5 \\ \hline 23 \\ 20 \\ \hline 3 \end{array}$$

$$\begin{array}{r} 14 \\ \times 5 \\ \hline 70 \\ + 3 \\ \hline 73 \end{array}$$

$$4 \overline{)94} \qquad \times \underline{}$$

$$2 \overline{)97} \qquad \times \underline{}$$

$$3 \overline{)82} \qquad \times \underline{}$$

$$7 \overline{)88} \qquad \times \underline{}$$

$$2 \overline{)71} \qquad \times \underline{}$$

$$6 \overline{)75} \qquad \times \underline{}$$

$$8 \overline{)99} \qquad \times \underline{}$$

DON'T WALK

76

Checking Division with Remainders

Solve.

1. Dennis has 45 pieces of wood. The wood stove burns 2 pieces of wood each hour. How many hours will the wood last? How many pieces will be left?

2. There are 66 feet of ribbon on a roll. Chris needs 4 feet of ribbon to wrap each gift. How many gifts can she wrap? How many feet of ribbon will be left?

3. Mark collected 75 soda bottles. 6 bottles fit in a pack. How many packs can Mark fill? How many bottles will he have left over?

4. Mrs. Fitz canned 97 jars of peaches. If she puts 8 jars on a shelf, how many shelves did she fill? How many jars did she have left?

5. There are 78 people who want to play cards. How many groups of 4 people can play? How many people cannot play?

6. Otis is taking 50 bricks to the backyard. If he carries 3 bricks at a time, how many trips will he make? How many bricks will he carry on the extra trip?

7. Mr. Ming owns 64 acres of land. He divided it into 3 equal parts. How many acres were in each part? How many acres were left?

8. Tammy has 27 yards of cloth. How many banners can she make if each is 2 yards long? How much cloth will she have left?

Name _____

Dividing hundreds is as easy as dividing ones and tens.
Remember to put 0s in the ones and tens places.

6 ones $\div 2 = 3$ ones	6 tens $\div 2 = 3$ tens	6 hundreds $\div 2 = 3$ hundreds
$6 \div 2 = 3$	$60 \div 2 = 30$	$600 \div 2 = 300$
$2\overline{)6}^{\,3}$	$2\overline{)60}^{\,30}$	$2\overline{)600}^{\,300}$

Divide.

$2\overline{)8}^{\,4}$ $2\overline{)800}^{\,400}$

$1\overline{)50}$ $1\overline{)500}$

$3\overline{)90}$ $3\overline{)900}$  $6\overline{)6}$ $6\overline{)600}$

$2\overline{)600}$ $3\overline{)30}$ $4\overline{)800}$ $8\overline{)80}$ $1\overline{)700}$

$2\overline{)200}$ $2\overline{)40}$ $3\overline{)600}$ $9\overline{)90}$ $2\overline{)400}$

$5\overline{)500}$ $3\overline{)90}$ $4\overline{)400}$ $1\overline{)70}$ $8\overline{)800}$

Dividing Hundreds
© The Continental Press, Inc.

Divide the hundreds. (How many 3s in 9?) Multiply. $3 \times 3 = 9$ Subtract. $9 - 9 = 0$	Bring down and divide the tens. (How many 3s in 6?) Multiply. $2 \times 3 = 6$ Subtract. $6 - 6 = 0$	Bring down and divide the ones. (How many 3s in 3?) Multiply. $1 \times 3 = 3$ Subtract. $3 - 3 = 0$
$\begin{array}{r} 3 \\ 3\overline{)963} \\ -9 \\ \hline 0 \end{array}$	$\begin{array}{r} 32 \\ 3\overline{)963} \\ -9 \\ \hline 06 \\ -6 \\ \hline 0 \end{array}$	$\begin{array}{r} 321 \\ 3\overline{)963} \\ -9 \\ \hline 06 \\ -6 \\ \hline 03 \\ -3 \\ \hline 0 \end{array}$

Divide.

$4\overline{)848}$ $2\overline{)284}$ $5\overline{)555}$ $3\overline{)639}$ $2\overline{)864}$

$3\overline{)996}$ $4\overline{)488}$ $2\overline{)886}$ $2\overline{)462}$ $3\overline{)393}$

$2\overline{)648}$ $3\overline{)963}$ $3\overline{)696}$ $4\overline{)884}$ $2\overline{)624}$

Name _____

Divide the hundreds.
(How many 2s in 9?)
Multiply. $4 \times 2 = 8$
Subtract. $9 - 8 = 1$

$$
\begin{array}{r}
4 \\
2\overline{)952} \\
-8 \\
\hline
1
\end{array}
$$

Bring down the tens
and regroup. Divide.
(How many 2s in 15?)
Multiply. $7 \times 2 = 14$
Subtract. $15 - 14 = 1$

$$
\begin{array}{r}
47 \\
2\overline{)952} \\
-8\downarrow \\
\hline
15 \\
-14 \\
\hline
1
\end{array}
$$

Bring down the ones
and regroup. Divide.
(How many 2s in 12?)
Multiply. $6 \times 2 = 12$
Subtract. $12 - 12 = 0$

$$
\begin{array}{r}
476 \\
2\overline{)952} \\
-8 \\
\hline
15 \\
-14\downarrow \\
\hline
12 \\
-12 \\
\hline
0
\end{array}
$$

Divide.

$3\overline{)822}$ $5\overline{)685}$ $2\overline{)716}$ $4\overline{)972}$

$6\overline{)756}$ $8\overline{)936}$ $3\overline{)768}$ $5\overline{)875}$

$2\overline{)992}$ $5\overline{)765}$ $3\overline{)864}$ $4\overline{)696}$

Dividing by 1 Digit for 3-Digit Quotients, with Regrouping

$$4\overline{)744}\quad\begin{array}{r}186\\\hline\end{array}$$

```
      186
   4 ) 744
      -4↓|
      ‾‾‾‾
       34|
      -32↓
      ‾‾‾‾
        24
       -24
      ‾‾‾‾
         0
```

Divide the hundreds.
Multiply. Then subtract.
Bring down the tens and regroup.
Divide the tens.
Multiply. Then subtract.
Bring down the ones and regroup.
Divide the ones.
Multiply. Then subtract.

Divide.

```
      154
   5 ) 770
       5
      ‾‾
      27
      25
      ‾‾
       20
       20
      ‾‾
```

$$7\overline{)896}\qquad 2\overline{)534}\qquad 6\overline{)918}$$

$$3\overline{)747}\qquad 8\overline{)928}\qquad 4\overline{)736}\qquad 2\overline{)918}$$

$$4\overline{)980}\qquad 5\overline{)925}\qquad 3\overline{)891}\qquad 6\overline{)702}$$

Dividing by 1 Digit for 3-Digit Quotients, with Regrouping
© The Continental Press, Inc.

Name _____

```
        165 R2
    5 ) 827
      - 5↓|
        32|
      -30↓
        27
      -25
         2
```

Divide the hundreds. $8 \div 5 = 1$
Multiply. Then subtract. $8 - 5 = 3$
Bring down the tens. Regroup if necessary.
Divide the tens. $32 \div 5 = 6$
Multiply. Then subtract. $32 - 30 = 2$
Bring down the ones. Regroup if necessary.
Divide the ones. $27 \div 5 = 5$
Multiply. Then subtract. $27 - 25 = 2$
Write the remainder R2 in the quotient.

Divide.

```
  324 R2
3 ) 974
    9
    7
    6
   14
   12
    2
```
6) 747   2) 455   8) 989

5) 591   4) 859   7) 905   3) 728

3) 671   4) 719   2) 937   5) 689

Dividing by 1 Digit for 3-Digit Quotients, with Regrouping and Remainders

Name _____

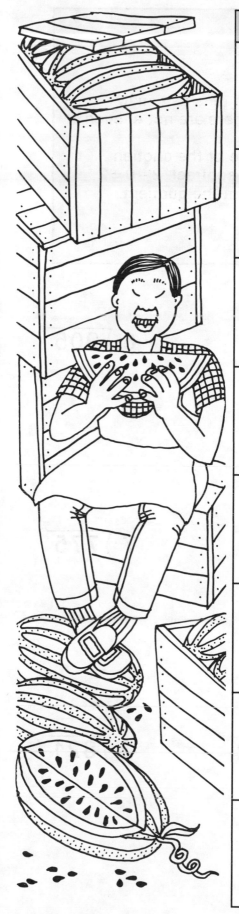

Solve.

1. The Valley Hotel has 369 rooms for guests. It has 3 floors with the same number of rooms per floor. How many rooms are on each floor?

2. There are 685 links in a chain that is 5 feet long. How many links are there in a foot?

3. This week 744 people visited a health center. If 4 doctors share the work, about how many people did each doctor see?

4. Mr. Nishito has 926 watermelons to ship. If 6 watermelons fit in a crate, how many crates will he need? How many watermelons will be left?

5. There are 424 girls at a summer camp. If 2 girls share a tent, how many tents are set up?

6. Last year Laurel's chickens laid 738 eggs. If she had 3 chickens, about how many eggs did each one lay?

7. At a supermarket, 663 ears of corn were wrapped in packages of 4 ears. How many packages were wrapped? How many ears of corn were left?

8. An artist printed 875 color posters in 7 days. How many did he print each day?

Name _____

Sometimes there are not enough ones to divide.

```
      250 R2
   3 ) 752
     -6↓ |
      15 |
     -15↓|
       02
      - 0
        2
```

Divide the hundreds and tens as usual.
Bring down the ones. There are not enough ones to divide.
Put a 0 in the ones place of the quotient.
Multiply. $0 \times 3 = 0$. Then subtract. $2 - 0 = 2$
Write the remainder R2 in the quotient.

Divide.

```
      140 R1
   5 ) 701
      5
      20
      20
       01
        0
        1
```

7) 846

3) 961

6) 905

8) 886

2) 781

5) 604

7) 775

4) 923

9) 998

3) 842

2) 821

84

Sometimes there are not enough tens to divide.

```
      109 R3
   4 ) 439
     -4↓|
       3 |
      -0↓
       39
      -36
        3
```

Divide the hundreds as usual.
Bring down the tens. There are not enough tens to divide.
Put a 0 in the tens place of the quotient.
Multiply. $0 \times 4 = 0$. Then subtract. $3 - 0 = 3$
Bring down the ones and regroup.
Divide the ones as usual.

Divide.

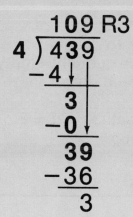

6) 642

5) 508

3) 923

8) 840

2) 613

7) 759

4) 827

2) 812

3) 626

9) 929

6) 656

4) 820

Sometimes there are not enough hundreds to divide.

```
      76 R1
4 ) 305
   -28↓
     25
    -24
      1
```

There are not enough hundreds to divide.
Leave the hundreds place of the quotient empty.
Regroup the hundreds as tens and divide.
(How many 4s in 30?)
Multiply. $7 \times 4 = 28$. Then subtract. $30 - 28 = 2$
Bring down the ones and regroup.
Divide the ones as usual.

Divide.

```
    20 R5
6 ) 125
    12
    ---
     5
    ---
     5
```
　　　　8) 363　　　　3) 172　　　　5) 375

2) 113　　　　4) 390　　　　7) 518　　　　3) 250

5) 479　　　　9) 570　　　　8) 296　　　　6) 439

Dividing by 1 Digit, 3-Digit Dividends and 2-Digit Quotients

Name _____

Solve.

1. A perfume company sent 384 free samples to 6 department stores. How many samples were sent to each store?

2. The Van Etten family put 4 wagonloads of hay in their barn. If there were 420 bales in all, how many bales were in each load?

3. DeSoto's Garden Center is selling 963 tomato plants in boxes of 4 plants. How many boxes are there? How many extra plants are there?

4. A movie took 215 days to make. The company worked 7 days a week. How many full weeks did it take to make the movie? How many extra days did it take?

5. A farmer has 782 bags of weed killer. If he spreads 3 bags an acre, how many acres can he treat? How many bags will be left?

6. A scientist used 846 test tubes in an experiment that lasted 9 weeks. If she used the same number each week, how many test tubes did she use in a week?

7. Sal copied a report that was 8 pages long. She ran off 872 pages on the copying machine. How many copies of each page did she make?

8. Mr. Baker drove 250 miles in 5 hours. How many miles did he drive an hour?

An estimated answer is a guess near the actual answer. To estimate the quotient, round the number being divided to the nearest ten or hundred. Then divide the rounded number.

Estimate

$$8 \overline{)39} \longrightarrow \overset{5}{8 \overline{)40}}$$

Estimate

$$2 \overline{)195} \longrightarrow \overset{100}{2 \overline{)200}}$$

Round to the nearest ten and estimate the quotient.

Round to the nearest hundred and estimate the quotient.

Estimate

$$3 \overline{)87} \longrightarrow \overset{30}{3 \overline{)90}}$$

Estimate

$$8 \overline{)375} \longrightarrow$$

Estimate

$$6 \overline{)25} \longrightarrow$$

Estimate

$$5 \overline{)210} \longrightarrow$$

Estimate

$$5 \overline{)58} \longrightarrow$$

Estimate

$$4 \overline{)750} \longrightarrow$$

Estimate

$$7 \overline{)74} \longrightarrow$$

Estimate

$$3 \overline{)944} \longrightarrow$$

Now divide the actual numbers. Compare your estimated quotient with the actual quotient.

Estimating Quotients
© The Continental Press, Inc.

Name _____

| Estimate each answer. Then solve. |

1. Tanya wrote 32 thank-you notes. She wrote 5 an hour. How many full hours did she work? How many notes were left?

2. A gardener planted 184 ivy plants in 4 hours. How many did she plant each hour?

3. Martin spent 71 hours of his vacation building a boat. If his vacation was 7 days long, about how many hours did he work on the boat each day?

4. There were 4 cakes at a party. If 84 people were served, how many pieces was each cake cut into?

5. Mr. Waltz bought a daily newspaper 312 times this year. If he bought it 6 days a week, how many weeks did he get it?

6. Several bike travelers rode 658 miles in 7 days. How many miles did they travel a day?

7. There are 45 puppets in a TV show. 5 people work equal numbers of them. How many puppets does each person work?

8. A metalworker stamped out 624 parts today. If he worked 8 hours, how many parts did he stamp out each hour?

Name _____

Divide dollars and cents the same way
you divide other numbers. Remember
to write the dollar sign and decimal
point in the answer.

$$
\begin{array}{r}
\$1.09 \\
7 \overline{)\ \$7.63} \\
-7 \\
\hline
6 \\
-0 \\
\hline
63 \\
-63 \\
\hline
0
\end{array}
$$

Divide.

$9 \overline{)\ \$9.18}$ $3 \overline{)\ \$6.96}$ $5 \overline{)\ \$7.25}$ $2 \overline{)\ \$8.42}$

$4 \overline{)\ \$5.08}$ $6 \overline{)\ \$8.70}$ $8 \overline{)\ \$5.36}$ $7 \overline{)\ \$6.09}$

$5 \overline{)\ \$2.95}$ $9 \overline{)\ \$8.46}$ $4 \overline{)\ \$3.60}$ $6 \overline{)\ \$5.64}$

Name _____

Find the average of each group of numbers.

```
   9        1 2           33                    $1.34
+ 15     2 ) 2 4        + 27        )          + 2.46        )
  24         2
         -------
            4
            4
         -------
```

```
   4                    $.12                     359
   8        )            .45        )            508          )
+  6                   + .15                    + 93
```

```
$.07                     17                      207
 .08        )            20        )             101          )
 .08                     41                      431
+.09                   + 14                    + 125
```

Name _____

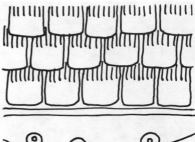

Solve.

1. Wayne read 42 pages today and 34 pages yesterday. What is the average number of pages he read a day?

2. Tablets at Thrift Mart are priced at $.49, $.67, and $.76. What is the average price of a tablet?

3. Jess bought 3 magazines for $.75, $1.25, and $2.50. What was the average price Jess paid for a magazine?

4. Exton had 5 inches of snow in December, 9 inches in January, 22 inches in February, and 16 inches in March. What was the average snowfall a month?

5. On a trip, Amy drove 85 miles, José drove 102 miles, and Erica drove 125 miles. What was the average distance driven?

6. Frank bought 2 tapes for $4.24 and $5.10. What was the average price of a tape?

7. Mrs. Vargas made 4 phone calls that were 35, 22, 21, and 18 minutes long. What was the average length of a call?

8. Regular gas costs $1.28 a gallon. Unleaded costs $1.37 and super costs $1.43. What is the average price of a gallon of gas?

Problem Solving: Finding Averages
© The Continental Press, Inc.

Name _____

Divide to find the cost of one item that is priced as part of a group.
The / sign means "for," as in "6 cans for $1.20."

$$6 \overline{)\ \$1.20} \quad \$\ .20 \text{ each can}$$
$$\underline{1\ 20}$$

Divide to find the cost of one of each group of items or the price of a pound.

2 packs ice cream / $1.80	6 cans orange juice / $3.00	4 light bulbs / $2.52
$$2 \overline{)\ \$1.80} \quad \$.90$$ $$\underline{1\ 80}$$		
2 rolls paper towels / $1.42	3 packs fish sticks / $2.40	5 pounds of onions / $1.45
8 packs dessert mix / $2.64	7 doughnuts / $2.10	9 pounds of grapes / $6.75

Add, subtract, multiply, or divide to solve each problem.

1. Judy bought a sack of whole wheat flour for $1.20 and some honey for $2.10. How much did she spend in all?

2. Every hour 15 cars can go through a car wash. How many cars can go through in 8 hours?

3. A gift shop had 560 funny bumper stickers. It sold 288 of them in a week. How many were left?

4. Monica had 112 comic books. Her uncle gave her 89 old comic books he found in his attic. How many comic books does Monica have now?

5. Dave attended an auto mechanic school for 360 days. He went to class 5 days a week. How many weeks did Dave attend the school?

6. Today turkey costs $.92 a pound. How much does a 9-pound turkey cost?

7. A sleeping bag is on sale for $76.95. Its regular price is $95.50. How much cheaper is it on sale?

8. A 7-day bus ticket costs $4.55. How much does it cost a day?

Problem Solving: Choosing Among Four Operations

Name _____

HAMBURGER $2.10
CHEESEBURGER $2.25
DOUBLE-BURGER $2.68
MUSHROOM BURGER $2.72
BELTBUSTER $3.50
FRENCH FRIES . . . . $.75
ONION RINGS . . . . $.75
SODA . . . . . . small $.40, large $.70
MILK SHAKE . . . . $.92

Read the menu. Then add, subtract, multiply, or divide to solve the problems below.

1. Curtis ordered a hamburger and french fries. What was his bill?

2. Jill has $3.00. If she orders a mushroom burger, how much money will she have left?

3. Leila bought onion rings. How much change did she get from $1.00?

4. Dale, Megan, and Glen ordered cheeseburgers. What was their bill?

5. Roger had a Beltbuster, onion rings, and a milk shake. What was his bill?

5. How much less does a double-burger cost than a Beltbuster?

7. How much do 7 orders of french fries cost?

8. Twins Sal and Val shared a Beltbuster. How much did each girl's share cost?

Problem Solving: Using a Menu
© The Continental Press, Inc.

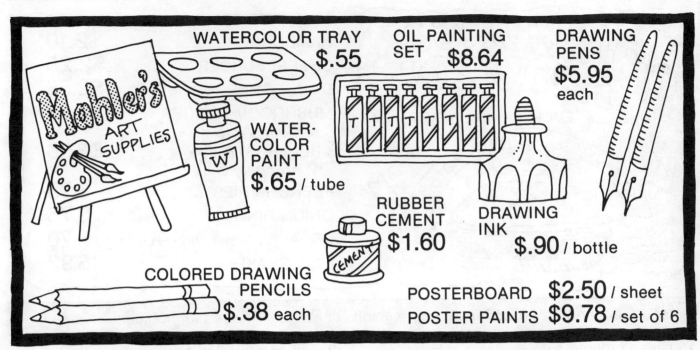

MOHLER'S ART SUPPLIES	
WATERCOLOR TRAY $.55	OIL PAINTING SET $8.64
WATER-COLOR PAINT $.65 / tube	DRAWING PENS $5.95 each
RUBBER CEMENT $1.60	DRAWING INK $.90 / bottle
COLORED DRAWING PENCILS $.38 each	POSTERBOARD $2.50 / sheet POSTER PAINTS $9.78 / set of 6

Read the ad. Then add, subtract, multiply, or divide to solve the problems.

1. Elaine bought a drawing pen and a bottle of drawing ink. How much did she spend in all?	**2.** Andy bought a plastic watercolor paint tray. How much change did he get from $1.00?
3. Kyoko bought 7 colored drawing pencils. How much did they cost in all?	**4.** How much does 1 bottle of poster paint cost?
5. Ms. Links bought an oil painting set. How much change did she get from $20.00?	**6.** What was Drew's bill for 9 tubes of watercolor paint?
7. Luis bought a sheet of posterboard, a set of poster paints, and a bottle of rubber cement. What was the total bill?	**8.** There are 8 tubes of oil paint in the set. How much does 1 tube cost?

96

Problem Solving: Using an Ad